Sustainable Packaging Symposium 2011

Chicago, Illinois, USA
16-18 March 2011

ISBN: 978-1-5108-1736-4

Printed from e-media with permission by:

Curran Associates, Inc.
57 Morehouse Lane
Red Hook, NY 12571

Some format issues inherent in the e-media version may also appear in this print version.

Printed by Curran Associates, Inc. (2016)

AIChE
120 Wall Street, FL 23
New York, NY 10005-4020

Phone: (800) 242-4363
Fax: (203) 775-5177

www.aiche.org

Additional copies of this publication are available from:

Curran Associates, Inc.
57 Morehouse Lane
Red Hook, NY 12571 USA
Phone: 845-758-0400
Fax: 845-758-2634
Email: curran@proceedings.com
Web: www.proceedings.com

TABLE OF CONTENTS

KEYNOTE DAY 1

SUPPLY CHAIN: RAW MATERIAL TO PACKAGE ON THE SHELF

HORIZON ISSUES: ON THE SUPPLY CHAIN

KEYNOTE DAY 2

MEASUREMENTS, DATA AND ANALYSIS

UNINTENDED CONSEQUENCES/GLOBAL PERSPECTIVE

YOUR M&S | FOODS PACKAGING

The Sustainable Future

Andrew Speck
Packaging Buyer

1

Business Overview

YOUR M&S | FOODS PACKAGING

- **Over £9bn sales annually**
- **Over 600 Stores in the UK**
- **Main activities are Clothing, Homeware and Foods**
- **Around 6% share of UK Food Market**
- **International Business**
- **Around 16 Million visitors to our stores each week.**

2

Simply Food

YOUR M&S | FOODS PACKAGING

BP Franchise

YOUR M&S | FOODS PACKAGING

Modernisation Programme

YOUR M&S | FOODS PACKAGING

New Initiatives

YOUR M&S | FOODS PACKAGING

8

Our Journey

YOUR M&S | FOODS PACKAGING

- **Marks & Spencer and Plan A**
- **Achievements to Date**
- **The Perfect Storm**
- **Plan A Phase 2**
- **The Sustainable Future**

11

Plan A

YOUR M&S

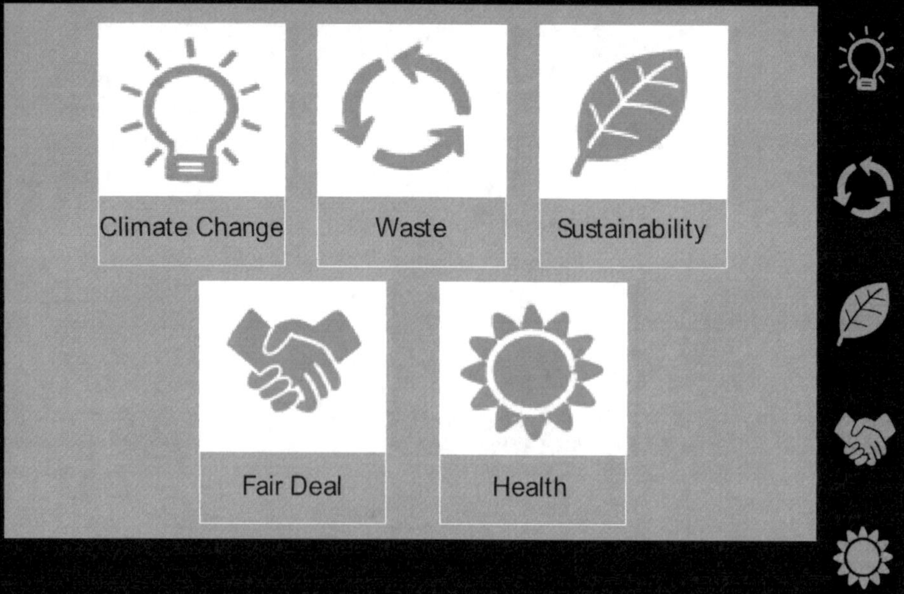

Plan A - Waste
YOUR M&S

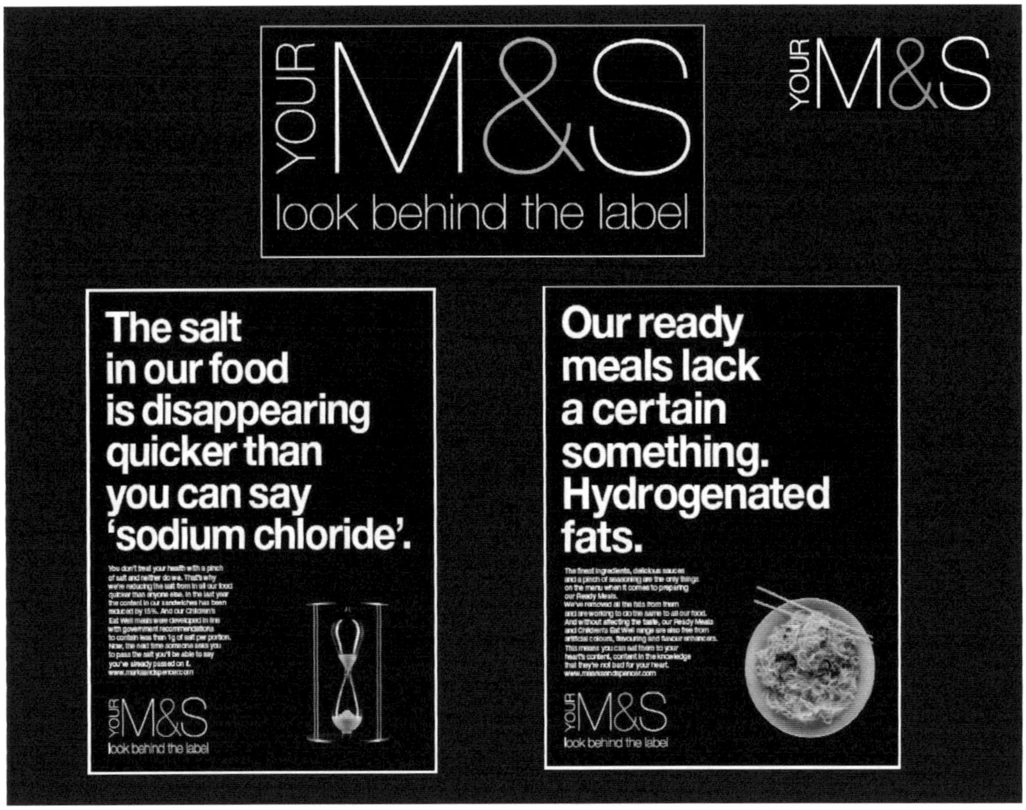
YOUR M&S
look behind the label
YOUR M&S
The salt in our food is disappearing quicker than you can say 'sodium chloride'.
YOUR M&S
look behind the label
Our ready meals lack a certain something. Hydrogenated fats.
YOUR M&S
look behind the label

Plan A - Raw Materials
YOUR M&S

Plan A – Fair Partner
YOUR M&S

YOUR M&S | FOODS PACKAGING

Video Presentation in Original e-media

17

Plan A Packaging Commitments

YOUR M&S | FOODS PACKAGING

- Reduction of packaging by 25%
- 100% of our packaging will be easily recyclable or compostable by the customer
- Increase the sustainability of packaging raw materials
- Labelling all products to help the customer dispose of their packaging in a responsible way

18

Recycling the Milk Bottle - Video

YOUR M&S | FOODS PACKAGING

Video Presentation in Original e-media

51

Plan A

Marks & Spencer Model Forest Programme

- Developing assessment of current sources for true sustainability
- Assessing existing suppliers
- Communicating programme to consumers
- Rewarding partners who are doing the right thing

Plan A DOING THE RIGHT THING

20% Packaging Reduction

YOUR M&S | FOODS PACKAGING

21

20% Packaging Reduction

YOUR M&S | FOODS PACKAGING

22

20% Packaging Reduction

YOUR M&S | FOODS PACKAGING

20% Packaging Reduction

YOUR M&S | FOODS PACKAGING

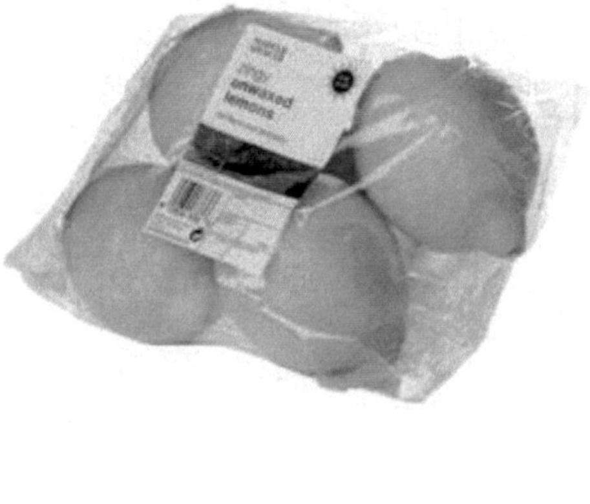

24

Reduction and Life Extension

YOUR M&S | FOODS PACKAGING

25

20% Packaging Reduction - Gifting

YOUR M&S | FOODS PACKAGING

26

20% Packaging Reduction

YOUR M&S FOODS PACKAGING

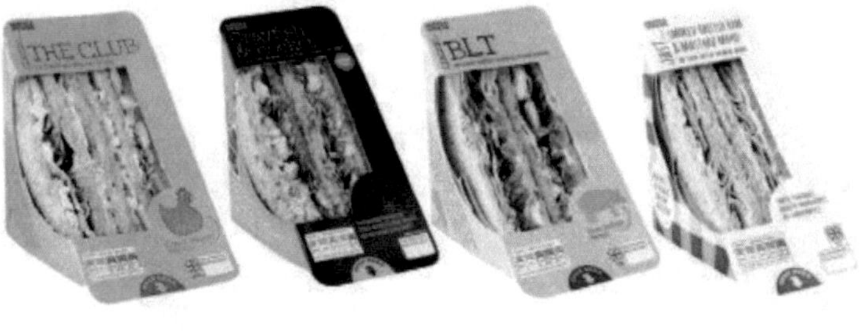

27

Recyclability of Packaging

YOUR M&S FOODS PACKAGING

'widely recycled'		'check local recycling'		'not currently recycled'	
e.g. format	material	e.g. format	material	e.g. format	material
Cereal carton/ tissue box	Cartonboard	Metal trays	Aluminium	Cling film/ Shrink wrap	PE
Drinks multi-pack	Corrugated cardboard	Plastic ready meal	C-PET	Plastic bags	LDPE
Egg box or fruit punnet	Moulded paper pulp	trays		Salad bags	OPP
Food and beverage cans	Aluminium and steel	Plastic salad bowls	A-PET	Pet food pouches	Plastic laminate
Aerosols	Aluminium and steel	Plastic fruit tray	PP / PET	Crisp packets	Plastic laminate
Wine bottles/ Jam jars	Glass	Plastic meat tray	PP / PET	Compostable bottles	PLA
Soft drinks bottles	Clear PET	Margarine tubs	PP / PS		
Milk bottles	HDPE	Soup/ sauce pots	PP		
Household cleaners	HDPE	Yogurt pot	PP / PS		
Drinks cartons	Composite, e.g. Tetra				
Total % by Weight	68%	Total % by Weight	23%	Total % by Weight	9%

74% 17% 9%

28

Sustainable Raw Materials

Recycled Content
Plastics

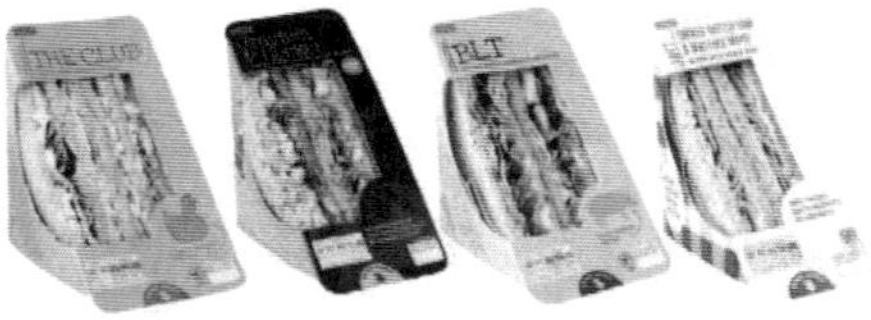

Sustainable Board

Food Waste

YOUR M&S FOODS PACKAGING

- Building customer awareness and knowledge
- All 7 depots now backhaul food waste from stores
- First step is MBT
- Next steps to progress to most environmentally beneficial solution

Out of the estimated 1 Million Tonnes per year of food waste from the retail sector, M&S accounts for around 36,000T from 600 shops – the challenge is to get this material, often in packaging, recovered in facilities in an economic way

What More Are We Doing?

YOUR M&S FOODS PACKAGING

- Design to Recycle
- Three Plastics
- Labelling with Recycling Logos
- Create an End-Market for Materials
- COMMUNICATIONS ARE KEY

SOURCES		DISPOSAL		
SLEEVE	SLEEVE	SLEEVE	TRAY	FILM
50% RECYCLED PAPER	FSC	paper	plastics	plastics
MINIMUM	MIXED SOURCES	WIDELY RECYCLED	CHECK LOCAL RECYCLING	NOT CURRENTLY RECYCLABLE

31

Communications

YOUR M&S FOODS PACKAGING

Melissavest.com

Stylehive.com

Ecocraftiness.com

Dogeared.com

Letting the story lead the facts

32

What's Next?

The 'Perfect Storm'

Growing populations, falling energy sources and food shortages will create the "perfect storm" by 2030,

The UK Government's Chief Scientist Professor John Beddington

The general challenges to this statement are around the timescales – not the theory itself

The world's rising population, 1950-2050

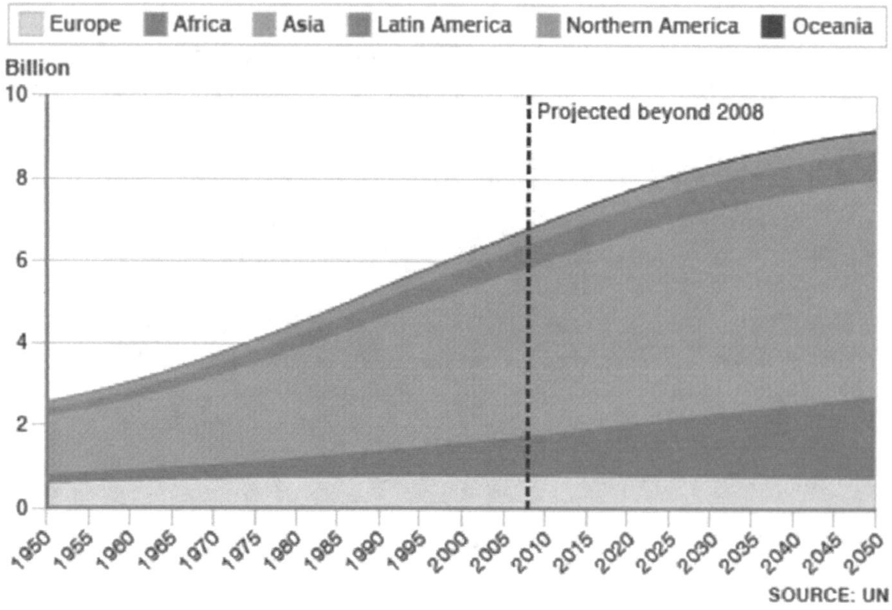

Water Availability 1961-90

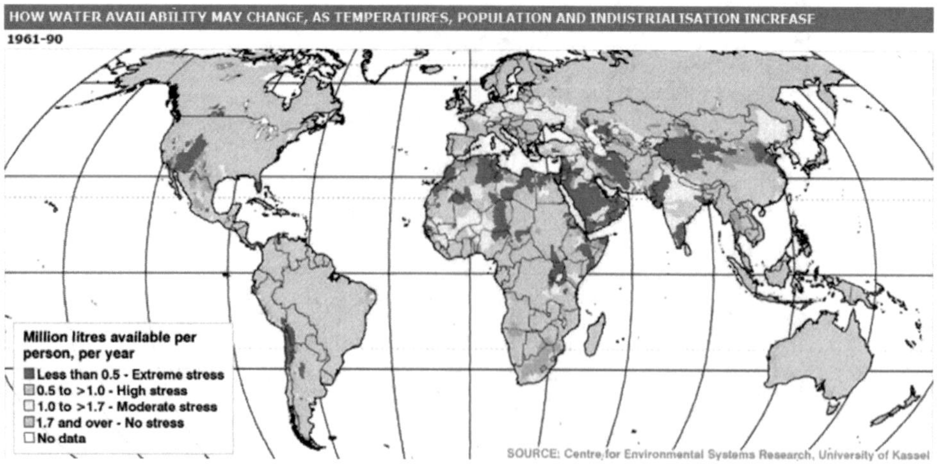

Water Availability 2020 Prediction

YOUR M&S | FOODS PACKAGING

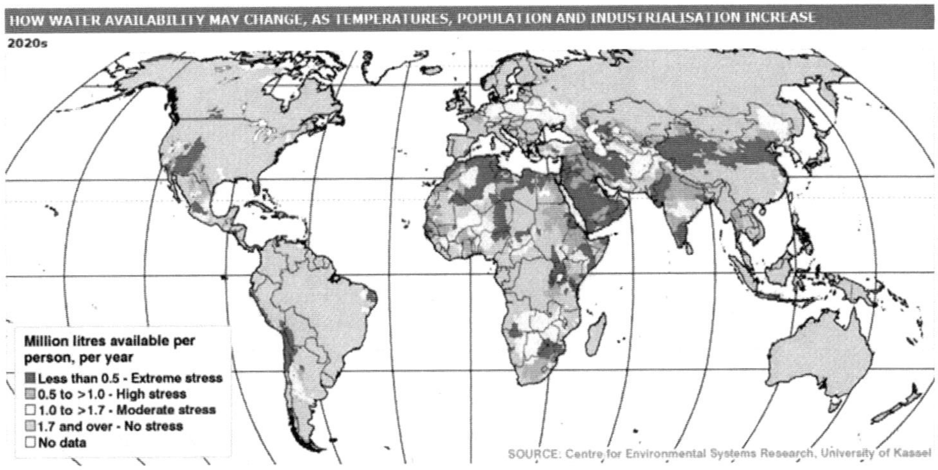

37

Water Availability 2050 Prediction

YOUR M&S | FOODS PACKAGING

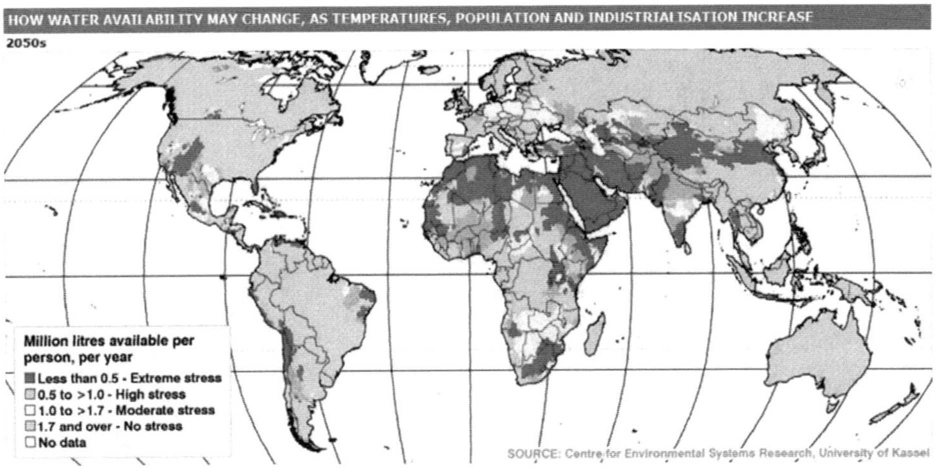

Centre for Environmental Systems Research, via BBC

38

How this has shaped our thinking

YOUR M&S FOODS PACKAGING

Simplistically the challenge for the future is:

- **Feed more people**
- **From the same amount of land**
- **Using half the amount of energy**

We have to act now

39

Our food business carbon footprint

YOUR M&S FOODS PACKAGING

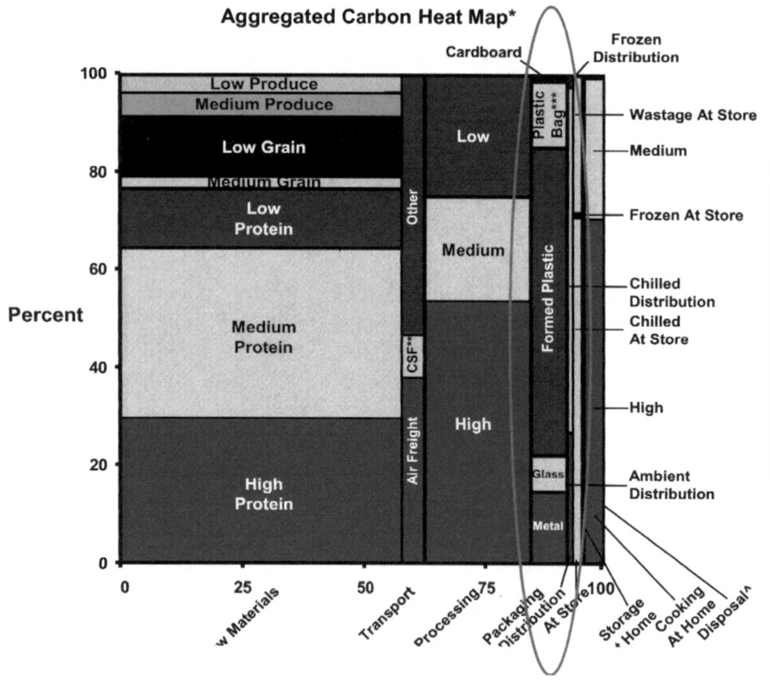

3.3m tonnes of CO2e per year

40

Plan A

Plan A – Phase 2

- Launched 3 years into Plan A phase 1
- Introduces further stretched goals for M&S
- Reflects greater understanding of environmental challenges
- Includes 35% reduction in carbon footprint of products by 2015
- Working towards the 'Sustainable Future'

Plan A DOING THE RIGHT THING

Plan A

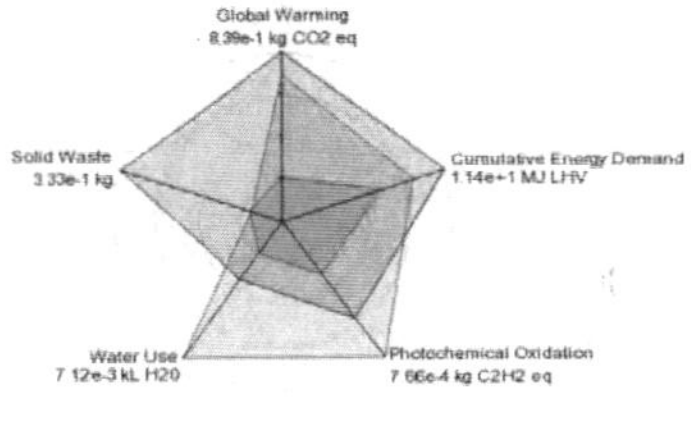

Somerset Waste Partnership

Plan A DOING THE RIGHT THING

Plan A Optimisation

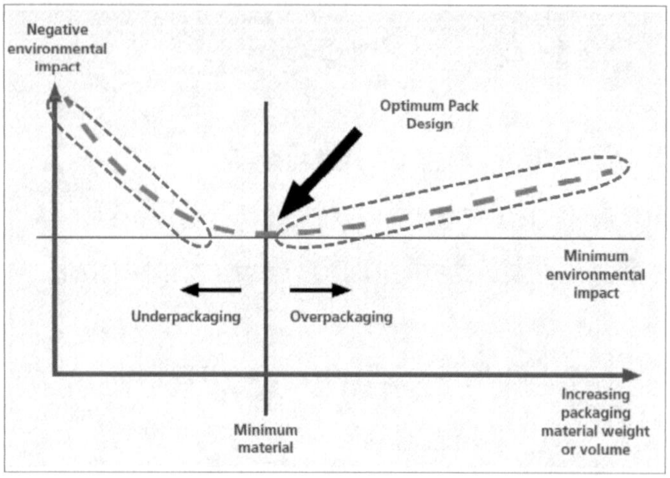

Optimum Packaging: The Innventia AB model shows that the environmental consequences of product losses caused by excessive packaging reduction are far greater than guaranteeing adequate protection through an incremental excess of packaging.

DOING THE RIGHT THING

Plan A Optimisation

supply chain enhancement

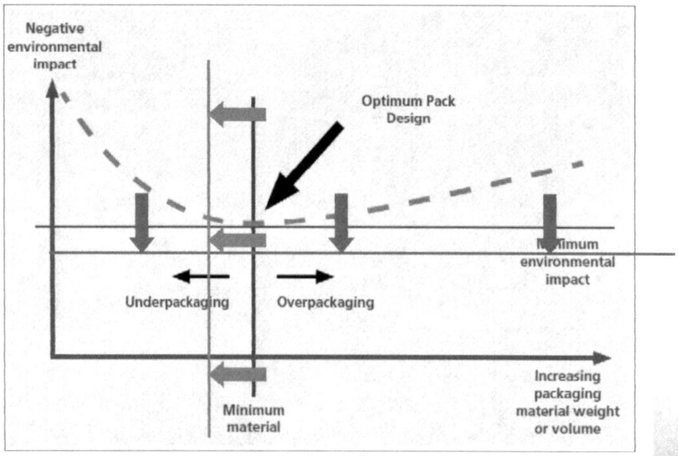

Optimum Packaging: The Innventia AB model shows that the environmental consequences of product losses caused by excessive packaging reduction are far greater than guaranteeing adequate protection through an incremental excess of packaging.

DOING THE RIGHT THING

Plan A Packaging Assessment Tool

PIQET FEATURES

- ✓ Identify ‘hotspots’
- ✓ Optimise design
- ✓ Rapid sensitivity analysis

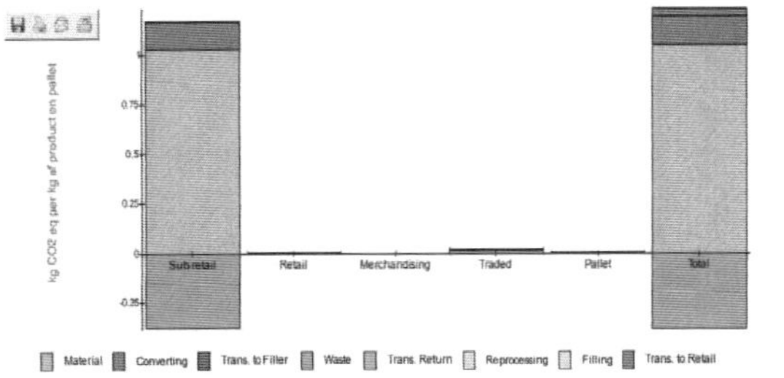

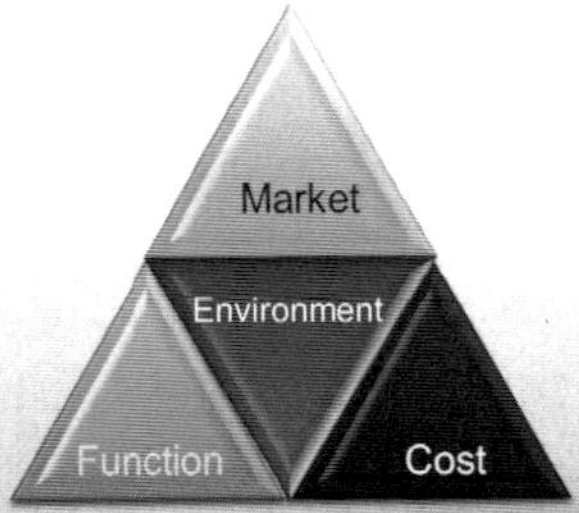

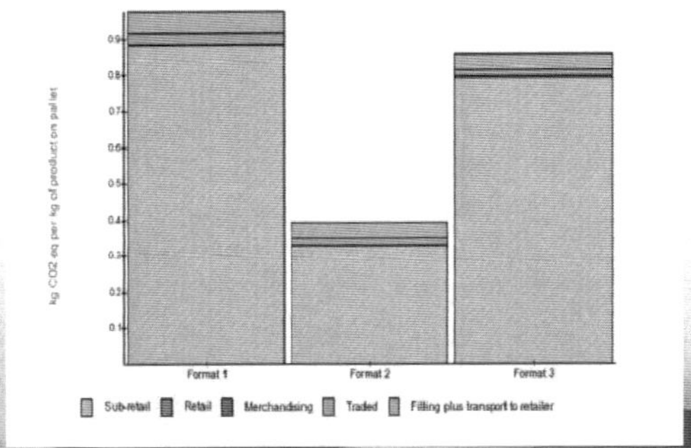

DOING THE RIGHT THING

Plan A

- ✓ Communicate
- ✓ Inform strategy
- ✓ Report
- ✓ Significant GPP compliance

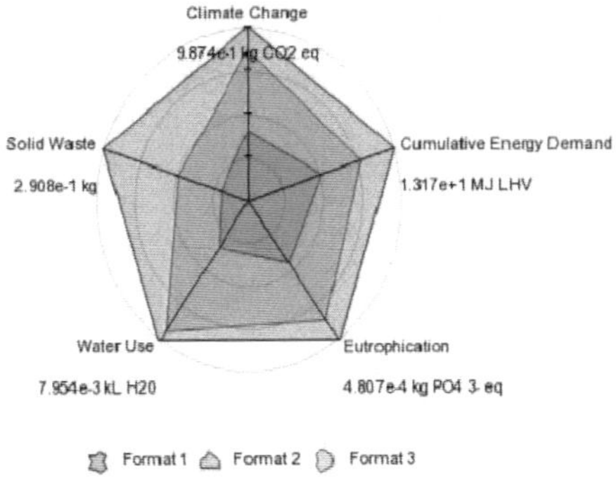

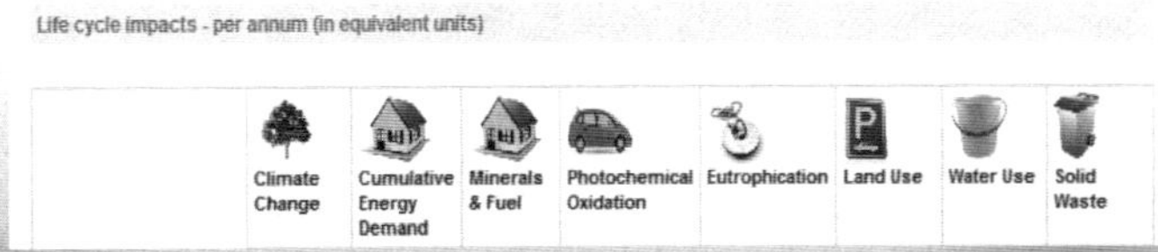

DOING THE RIGHT THING

Barriers to Recycling

YOUR M&S FOODS PACKAGING

Green Crusaders 30%

If it's easy 41%

What's the point? 23%

Not my problem 6%

WRAP report 'Barriers to Recycling'
www.wrap.org.uk/barrierstorecycling

47

How to maximise Recycling

YOUR M&S FOODS PACKAGING

Understanding our customers

	Incompetent	**Competent**
Conscious	***Level 2*** – consciously incompetent	***Level 3*** – consciously competent
Unconscious	***Level 1*** – unconsciously incompetent	***Level 4*** – unconsciously competent

Competency Learning Framework

Situational barriers:
Current recyclers say they would recycle a little or a lot more if they had:
#1 - Collections of a wider range of materials (52%).

Behaviour barriers:
Current recyclers nevertheless still sometimes or often:
#1 - Bin things because they are not sure if they can be recycled (48%).

Take away the barriers and the materials will flow

48

Recycling Partnerships

YOUR M&S | FOODS PACKAGING

We aim to work with within the local government structure to develop total recycling solutions for our customers across all packaging materials and formats that we use.

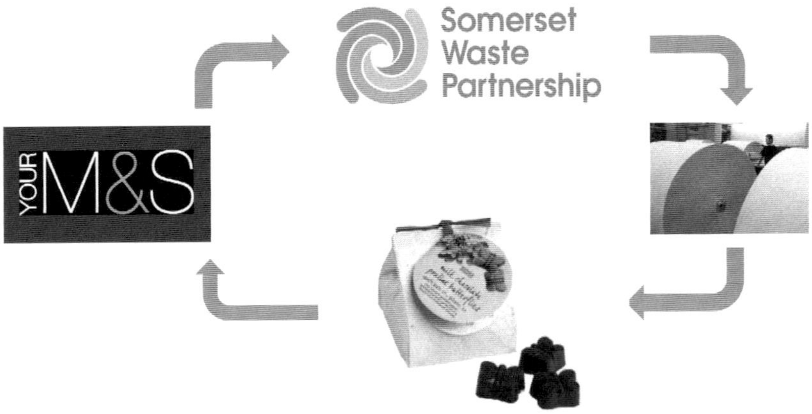

We have committed £1.25m over 5 years to roll our 'Sort It Plus' in Somerset

49

Local Authorities

YOUR M&S | FOODS PACKAGING

Connecting the last link in plastics value chain

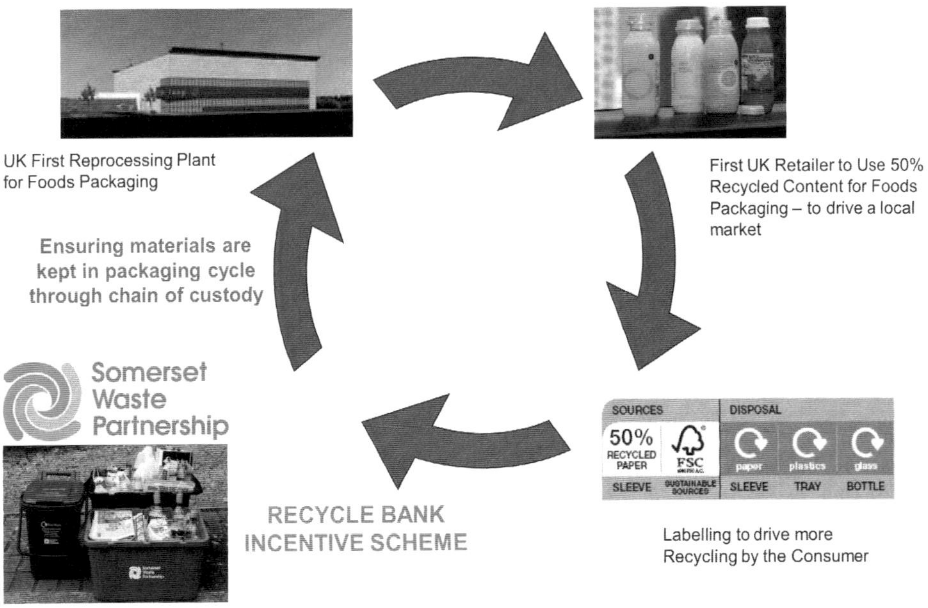

50

Recycling the Milk Bottle - Video

YOUR M&S | FOODS PACKAGING

Video Presentation in
Original e-media

51

Plan A

Marks & Spencer Model Forest Programme

- Developing assessment of current sources for true sustainability
- Assessing existing suppliers
- Communicating programme to consumers
- Rewarding partners who are doing the right thing

Plan A DOING THE RIGHT THING

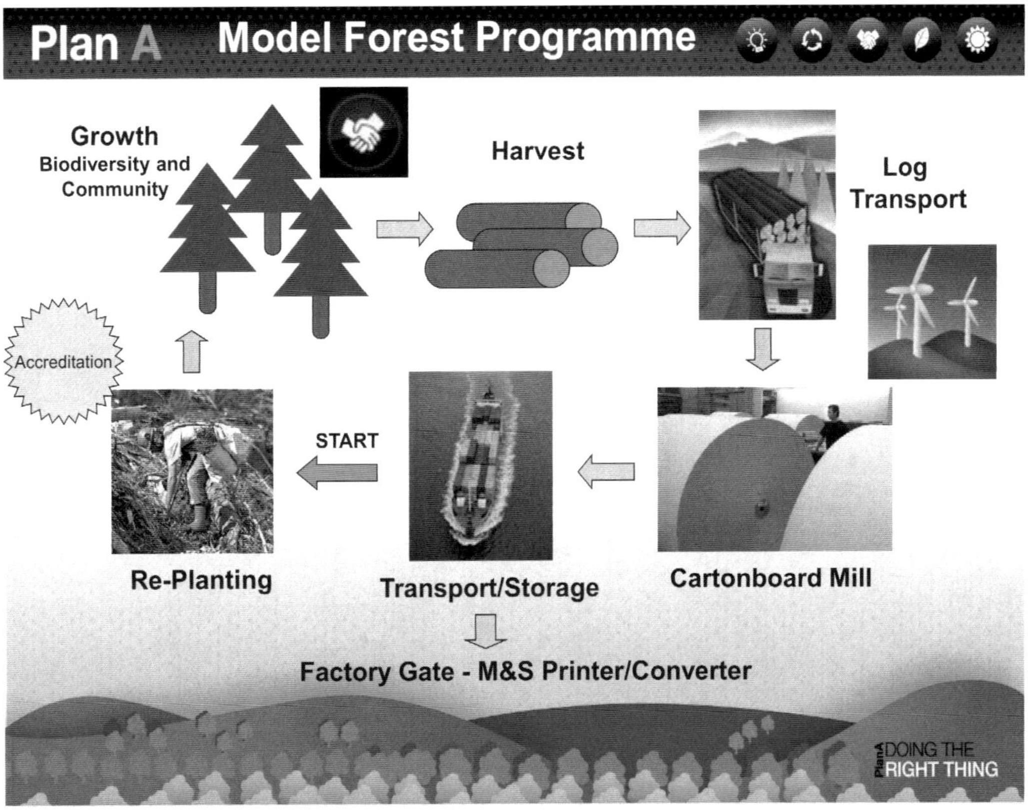

Communication

YOUR M&S FOODS PACKAGING

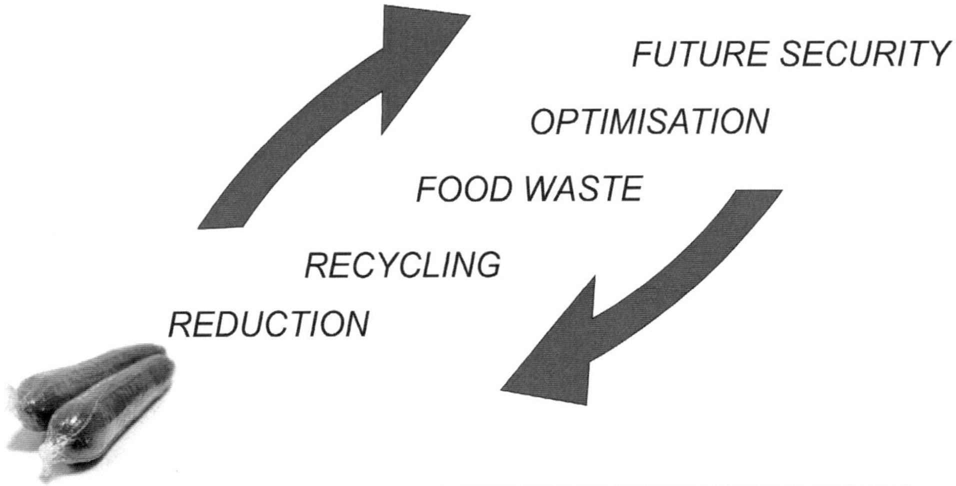

The journey can go quickly but public perception can constrain this thinking.
It is all our roles to educate and inform consistently and factually

54

Communication: Getting It Right

YOUR M&S | FOODS PACKAGING

55

But don't get Complacent

YOUR M&S | FOODS PACKAGING

56

Plan A
Plan A
The Sustainable Future
DOING THE
RIGHT THING

European Brands Association
des Industries de Marque
AIM®
Home
About AIM
Principles
The Brand
Policies
Insight & Innovation
Links
Member Zone
Program for Responsible Sourcing
AIM-PROGRESS is a forum of consumer goods companies assembled to enable and promote responsible sourcing practices and sustainable production systems. It is a global initiative supported and sponsored by AIM in Europe and the GMA in North America.
Its key objectives include the provision of a forum to exchange views regarding responsible sourcing practices and supporting the effective collaboration and potential convergence with other global initiatives having similar aims or interests.
It aims to develop and promote the use of common evaluation methods to determine CSR performance within the supply chain and drive efficiencies for all companies by collecting, assessing and sharing non-competitive information on supply chain CSR performance.
"Europe depends on powerful brands"
European Commission
Classified - Internal use
9 Avenue des Gaulois B-1040 Brussels Belgium :: Tel. +32 2 736 0305 :: Fax +32 2 734 6702 :: Email

Our Members
AIM®
Association des Industries de Marque
European Brands Association
Europäischer Markenverband
FERRERO
COLGATE-PALMOLIVE
DANONE
BACARDI
ABInBev
HERSHEY'S
The Hershey Company
Coca-Cola
DIAGEO
P&G
Kellogg's
Nestlé
Good Food, Good Life
BDF
Beiersdorf
Heineken
kraft foods
PEPSICO
MARS
incorporated
ORKLA
GENERAL MILLS
UB
Unilever
McCORMICK
Reckitt
Benckiser
BRITVIC
United Biscuits
Classified - Internal use

Responsible Sourcing

AIM®
Association des Industries de Marque
European Brands Association
Europäischer Markenverband

Branded goods manufacturers focus on building a sustainable business through sustainable communities includes Responsible Sourcing:

- Protect brand and company reputations from damage due to inappropriate practices against workers in the supply chain
- Take active measures identify and to efficiently mitigate such risks
- Increase supply chain performance / sustainability

Classified - Internal use

Our Members Approach

AIM®
Association des Industries de Marque
European Brands Association
Europäischer Markenverband

- **Establish standards such as Supplier Codes of Conduct**
- **Establish appropriate due diligence / monitoring of conditions through:**
  - Creation of policy and supporting management systems
  - "Lift and Shift" what others, such as Global Social Compliance Program, develop
  - Focus on effective deployment and not invention
  - Communicating and contracting requirements
  - Supplier self assessment and self certification
  - Supplier audits
  - Engage suppliers to correct issues detected
- **Find areas of pre-competitive joint interest to drive improvement:**
  - Systems such as SEDEX
  - Customer alignment such as Global Social Compliance Program (GSCP)
  - Common product areas
    - Agriculture with an initial focus on sugar
    - Promotional materials

Classified - Internal use

Aligned to the 10 Principles of the UN Global Compact

AIM®
Association des Industries de Marque
European Brands Association
Europäischer Markenverband

- **Human Rights and Labour Standards**
  - Defined by UNHR, ILO Conventions, local law and other norms
- **Health and Safety**
  - Defined by local law and buyer requirements
- **Environmental Compliance**
  - Concerned with site operations, i.e. water, energy, waste, and emissions controls, etc.
- **Business Integrity**
  - Policies for the mitigation of bribery & corruption, etc.

Classified - Internal use

Our Opportunity

AIM®
Association des Industries de Marque
European Brands Association
Europäischer Markenverband

- As individual companies we were asking our suppliers for similar assurances, but…
  - Used **different** words in **questions**,
  - Against **similar**, but different **standards**,
  - On **different systems**,
  - With **different audit** protocols
  - In **isolation** without finding synergy with like efforts
  - Leading to confusion and complexity and **duplication** of effort

Classified - Internal use

Our Objectives

AIM®
Association des Industries de Marque
European Brands Association
Europäischer Markenverband

- Provide a forum for developing common viewpoints and positions regarding responsible sourcing practices.
- Support the effective collaboration and potential alignment with other global initiatives having similar aims or interests such as GSCP.
- Develop and promote the use of common evaluation methods to determine performance within the supply chain.
- Drive efficiencies by collecting, assessing and sharing non-competitive information on supply chain performance.
- Share learning and information amongst Participating FMCGs, within the supply chain and its relevant stakeholders.

Classified - Internal use

Our Structure – Background

AIM®
Association des Industries de Marque
European Brands Association
Europäischer Markenverband

- Progress originally formed with guidance from CPO steering group
- Need for a link to NGOs and other external bodies for both ensuring the appropriate scope and deriving learning / credibility
- Need for a strong link to AIM Board and other taskforces
- Membership and scope growing
- Need identified to establish Taskforce Leadership team & Advisory Board to guide and manage activities

Classified - Internal use

Our Structure - Leadership Team

AIM®
Association des Industries de Marque
European Brands Association
Europäischer Markenverband

The key objectives of this group are;

- Provide direction to the taskforce
  - Ensure short and longer term objectives are set appropriate to priorities
  - Ensure workstreams are adequately set up, resourced and working effectively
  - Deliver taskforce group meetings with appropriate content and attendance
- Building the AIM-Progress Brand through creating communications and external engagement to raise profile and credibility to
- Ensuring the taskforce Charter and membership criteria are clearly set out, then driving recruitment and alignment with other AIM/external initiatives

Classified - Internal use

Our Structure - Leadership Team

AIM®
Association des Industries de Marque
European Brands Association
Europäischer Markenverband

David Lawrence	Diageo	Chair
Stuart Kyle	The Coca-Cola Company	Vice-Chair
Petar Sobic	Unilever	Past Chair
Russ Wood	PepsiCo	
Denise Wooldridge	Kraft Foods	
Maura Murphy	General Mills	
Benjamin Ware	Nestle	
Katrin Recke	AIM	Taskforce Secretariat

Classified - Internal use

Our Structure – Advisory Board

AIM®
Association des Industries de Marque
European Brands Association
Europäischer Markenverband

The key objectives of this group are;

- **Provide challenge/stimulation /advice to the taskforce leadership team**
  - Inspiring appropriate breadth/depth of scope to address collaborative opportunities, through bringing external perspectives and best practice
- **Be the link back in to the main AIM Board for support and alignment**
  - Ensuring no gaps or unnecessary overlaps with other taskforces
  - Drive increased membership of this taskforce with AIM members
- **Ensure the credibility for stakeholders of the taskforce work**

Classified - Internal use

Our Structure – Advisory Board

AIM®
Association des Industries de Marque
European Brands Association
Europäischer Markenverband

James Quincy	AIM Board Member	Chair and AIM Board sponsor
Steve Yucknut	Kraft Foods	Consumer Goods Forum link
Burkhard Gerstenkorn	Unilever	Procurement compliance
Saliha Barlatey	Nestle	Consumer Goods Forum link
Simon Steyne	ILO (IPEC)	Child Labor expertise
Sune Thorsen	Global CSR	Human Rights expertise
Patrick Laine	WWF	Environmental expertise
David Lawrence	AIM-Progress	Taskforce link (Chair)
Stuart Kyle	AIM-Progress	Taskforce link (Vice-chair)
Alain Galaski	AIM	AIM Link

Classified - Internal use

Draft Membership Criteria

AIM®
Association des Industries de Marque
European Brands Association
Europäischer Markenverband

- We are focused on our Mission in the FMCG sector
- We are open to all AIM and GMA members
- Being inclusive, we are open to any manufacturer and associated company integral to the FMCG manufacturer's supply chain that meets the 'expectations of the members'.

- **Expectations of members**
  - Support the Mission of the Taskforce
  - Attend at least one Taskforce meeting per year
  - Actively participate in at least one Taskforce workstream
  - Share current process details for benchmarking
  - Contribute to shared audited supplier list, where audits exist
  - Have some overlap of suppliers within the supply chain
  - Hold themselves to at least the same standards as they require of their suppliers
  - Be current on Taskforce fees

13

Classified - Internal use

The Work - Workstream Overview

AIM®
Association des Industries de Marque
European Brands Association
Europäischer Markenverband

Mutual Recognition	**"Done for one, done for all"** **Raising the bottom** **Reducing variability towards harmonization**
Capacity Building	**Awareness and training** **Remediation – performance improvement** **Commodity focus on Promotional Materials**
Systems	**SEDEX** **Other platforms** **Interoperability**
Agriculture / Extractives	**Sugar / Bonsucro** **Other agriculture** **Extractives**
Business Integrity	**Enabling practices / systems** **Corruption (fraud, bribery etc)** **Regulatory Requirements (FCPA, Anti-trust etc.)**

Classified - Internal use

Mutual Recognition Purpose and Goals

AIM®

Association des Industries de Marque
European Brands Association
Europäischer Markenverband

An Audit for One is an Audit For All

Purpose/Benefits

- Reduce cost burden on suppliers
- Move towards harmonization by sharing and adopt best practices
- Send a strong, unified message to suppliers
- Align with customers

2009-2011 Output

- Benchmarking completed & overlaps recognized
- Framework to mutually recognize assessments created and signed by 16 companies
- Enhance the work recognizing emerging external expectations

Classified - Internal use

Protocol Benchmarking

AIM®

Association des Industries de Marque
European Brands Association
Europäischer Markenverband

Row	AUDIT STANDARDS BENCHMARKING					
2	CODI CORE ELEMENTS COVERED	COMPANIES				
3						
5		Company 1	Company 2	SMETA	Company 3	Company 4
6 / 25	Forced Labor	X 100%	X 100%	X 100%	X 86%	X 71%
26 / 52	Wages & Benefits	X 90%	X 90%	X 70%	X 80%	X 60%
53 / 71	Hours of Work	X 100%	X 100%	X 100%	X 67%	X 67%
72 / 93	Freedom of Association	X 100%	X 50%	X 67%	X 67%	0%
94 / 113	Child Labor	X 67%	X 50%	X 100%	X 83%	X 50%
114 / 136	Discrimination	X 100%	X 57%	X 71%	X 100%	X 29%
137 / 153	Abuse of Labor	X 100%	X 80%	X 100%	X 100%	20%
154 / 201	Health & Safety	X 76%	X 71%	X 48%	X 76%	X 38%
202 / 277	Environment	X 78%	X 67%	X 56%	X 100%	0%
278 / 298	Business Integrity	X 44%	0%	X 44%	X 56%	0%
299 / 308	Regular Employment	100%	67%	X 100%	67%	67%
309						
310	OVERALL AVERAGE	87%	67%	78%	80%	36%

Baseline initially was SEDEX SMETA – but some protocols covered other areas.
Moved to more neutral approach, incorporating new developments from members.

Classified - Internal use

Additional Detail

AIM®
Association des Industries de Marque
European Brands Association
Europäischer Markenverband

	AUDIT STANDARDS BENCHMARKING					
2	CODI CORE ELEMENTS COVERED	COMPANIES				
3						
5		Company 1	Company 2	SMETA	Company 3	Company 4
6	Forced Labor	X	X	X	X	X
25		100%	100%	100%	86%	71%
26	Wages & Benefits	X	X	X	X	X
28	Local Law - Minimum Wage & OT	X	X	X	X	X
30	Payroll Records Maintained	X	X	X	X	X
32	Pay Slips Provided	X	X	X	X	X
34	Regular Payment	X	X	X	X	
36	Wage calculation (data collection: wage paid)	X	X		X	X
38	Deductions	X	X	X	X	
40	Benefits	X	X	X	X	X
42	Piece-Rate Calculation	X		X		X
44	Seasonal/Temporary Workers		X		X	
46	Training/Probationary Programs	X	X			
48	Other	Unpaid Meeting time				Workers surveys/OT knowledge
51						
52		90%	90%	70%	80%	60%
53	Hours of Work	X	X	X	X	X
71		100%	100%	100%	67%	67%
72	Freedom of Association	X	X	X	X	
93		100%	60%	67%	67%	0%

- Categories were included when three or more protocols referenced a specific topic. An "X" indicates that a company protocol references the sub-element in some way.
- When less than three protocols referenced a violation then the information is included under "other"

Benchmarking Summary

AIM®
Association des Industries de Marque
European Brands Association
Europäischer Markenverband

MUTUAL RECOGNITION BENCHMARKING SUMMARY

Results:
There is significant overlap in the protocols: there is ~78% overlap in the labor categories identified in the benchmarking exercise and 63% overlap in H&S and Environment categories

Methodology:
Categories were included when three or more protocols referenced a specific topic. An "X" indicates that a company protocol references the sub-element in some way. The detailed wording is included under each category and can be viewed by clicking the "+" on the left. When less than three protocols referenced a violation then the information is included under "other"

Company	How Many Pillars Covered?	% Benchmarking	Third Party Auditors Used
Company 1	4	75%	Not Always
Company 2	4	87%	Yes
Company 3	3	67%	Yes
Company 4	4	80%	Yes
Company 5	3	36%	Not Always
Company 6	4	78%	Not Always

Overview tab with results, methodology and summary

Mutual Recognition Framework

AIM®

Association des Industries de Marque
European Brands Association
Europäischer Markenverband

Mutual Recognition Criteria Proposal

Objective: Create a base criteria to facilitate companies acceptance and use of supplier audits completed on behalf of other companies while respecting code differences.
How: Collect and report relevant information to enable individual companies to verify audit integrity and determine compliance with regards to their respective internal standards.

Four Criteria:

- Acceptable Audit Coverage
- Acceptable Auditor
- Acceptable Process
- Audit Integrity

Relevant Information included on Audit Form
1. Date of audit
2. Auditor Firm Name; Auditor Name
3. Announced or unannounced
4. Number of man days spent at supplier
5. Sample size methodology used

I. Acceptable Audit Coverage
- Verify compliance against 10 key topic areas[1,2]
  - Compliance with local law
  - Environmental compliance
  - Freedom of association
  - Physical and mental abuse
  - Workers hours and overtime
  - Discrimination/harassment
  - Forced labor
  - Health and safety
  - Wages and benefits
  - Child Labor
- Audit is expected to verify compliance with local law in all subject areas
- Standard being verified is clearly stated so as to facilitate individual company decisions
- Key subject areas and data collection points are suggested in Appendix (based on benchmarking)

II. Acceptable Auditor
- Use of external auditors provides an assurance of objectivity which facilitates sharing among companies
- Recognized leading accrediting bodies for audit firms include[3]:
  - FLA, ICTI, WRAP, SAI

III. Audit Process
- The methodology used by an auditor to determine noncompliance is based on visual inspection, documentation review and interviews. The audit process includes, at a minimum, the following components:
  - Opening and closing meetings
  - Review of relevant documents
  - Facility inspection
  - Confidential employee interviews

Caveats
1. Companies will retain right to recognize (or not recognize) any audit or part of an audit
2. Companies will retain right to require additional followup
3. Companies not responsible for audits they sponsor

IV. Audit Integrity
- An acceptable audit report is complete and not redacted (except confidential production information)
- Audit firm and auditor name listed to facilitate verification
- Suggested means to obtain audit report is directly from auditor or by means of the Sedex data base or other such shared platform

[1] Model Reference Codes: BSCI, GSCP, SA8000, SEDEX (ETI)
[2] Supplier Codes of Conduct: Burger King, Coca-Cola, Con-Agra, Costco, Danone, Disney, Heinz, Loblaw, McDonald's, Nestle, Pepsico, Proctor & Gamble, Sara Lee, Sodexo, Sysco, Unilever, Wal-Mart, Wrigley, Yum!

- **4 criteria:**
  - Acceptable Audit Coverage
  - Acceptable Auditor
  - Acceptable Process
  - Audit Integrity
- **Need to update Auditor criteria**
  - GMI, GSCP learning

Classified - Internal use

Member Communication

AIM®

Association des Industries de Marque
European Brands Association
Europäischer Markenverband

AIM®

AIM-PROGRESS[1] MUTUAL RECOGNITION

TO REDUCE AUDIT FATIGUE AND ALIGN ON KEY ISSUES OF RESPONSIBLE SOURCING, SEVERAL BRAND MANUFACTURERS AGREED TO MUTUAL RECOGNITION OF SOCIAL COMPLIANCE ASSESSMENTS[2]

The companies listed below are part of a forum under AIM - the European Brands Association, called AIM-PROGRESS which seeks to promote responsible sourcing while reducing the duplication of supplier assessments. These companies have responsible sourcing programs which may include the request for on-site assessments of suppliers. The companies listed below agree to recognize supplier audits completed on behalf of another company and will review submitted assessment reports to confirm it meets internal company requirements.

Suppliers are therefore encouraged to share their assessment reports directly with these companies upon request or upload them via the Supplier Ethical Data Exchange (Sedex) platform, which provides a secure online data exchange between suppliers and customers. A supplier may only share assessment reports which they own the rights to or which they have received prior authorization to share.

Companies that agree to mutually recognize responsible sourcing assessments include:

AB Inbev
Bacardi
Britvic
Coca-Cola Enterprises
The Coca-Cola Company
Danone
Diageo
General Mills
Kellogg
Kraft Foods
Mars
Nestlé
Orkla
PepsiCo
Procter & Gamble
Unilever
United Biscuits

The companies listed here promote suppliers' sharing of assessment results while retaining the following rights.
1. To accept or not accept any audit or part of an audit
2. To require additional follow-up or conduct a full assessment
3. To verify the scope and integrity of a report. The audit report must have the auditor's name, affiliation and contact information. Other relevant information such as the standards used to measure compliance may be deemed necessary for a company to accept any report.

- Powerful communication tool with suppliers
- Updated version is always available online

Classified - Internal use

Capacity Building

AIM®

Association des Industries de Marque
European Brands Association
Europäischer Markenverband

2009 – 2010: Supplier Engagement and Employee Development

- **Over 2,000 people from over 1,000 companies participate**
  - Member hosted events in Turkey, Russia, India, China and Brazil
  - 4 Webcasts for North America and Europe
- **Target the deciders to get to yes and the doers on how to make it happen**
- **Attended by most senior member leaders in the geography**

2011: Supplier Engagement and Employee Development

- Target increased participation
- Member hosted events planned for Nigeria, Kenya, South Africa, Egypt, Mexico, Russia, and China
- Webcasts planned for the Americas (English and Spanish) and Europe
- Global Promotional Supplier Webcast
- **Added element of targeting advanced suppliers for a dialogue on capacity**

Classified - Internal use

Systems

AIM®

Association des Industries de Marque
European Brands Association
Europäischer Markenverband

Our Vision
Make Supply Chain Performance Visible from shelf to farm / point of extraction
Enabling
Visibility of gaps and deployment of appropriate resources to close gaps

Our Strategy
Deploy and Common Information Platform (SEDEX) for All Participants
Has Evolved
Expand Interoperability Catering to Additional Platforms

BUT OPERATING AS A VIRUTAL SINGLE PLATFORM

Our Approach
Leverage our Diversity to Capture Alignment, Innovation and Legacy

Classified - Internal use

Deploying Data Exchange to Drive Efficiencies

AIM®
Association des Industries de Marque
European Brands Association
Europäischer Markenverband

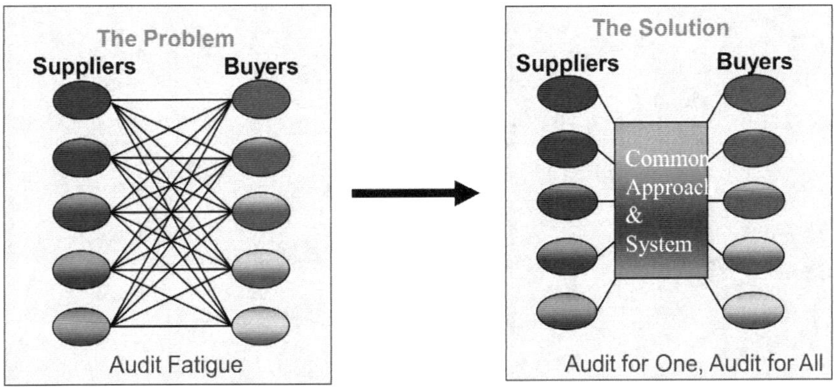

Classified - Internal use

System - the Value Proposition to AIM PROGRESS Members

AIM®
Association des Industries de Marque
European Brands Association
Europäischer Markenverband

- Data Exchange between Suppliers and Customers
- 'Farm (or mine) to Shelf' supply chain visibility
- One "Stop Shop" of Supplier Responsible Sourcing data
- Inclusive governance model
- Adoption of best practice tools
- Reporting Functionality
- Cost Efficient
- Interoperability with other Systems

Classified - Internal use

Summary of **The SEDEX System Assessment Process**

AIM® Association des Industries de Marque / European Brands Association / Europäischer Markenverband

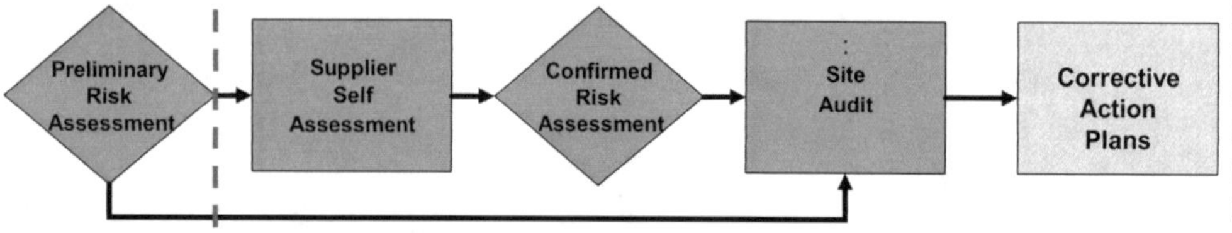

Independent Activity: Sedex Member determines which suppliers are requested to follow the assessment process.

Common Method: Supplier completes a site self-assessment questionnaire to a common format and is uploaded onto **Sedex database**

Independent Activity: Sedex Member determines which suppliers will be asked to perform an audit

Common Method: 3rd Party Auditor conducts a site audit to a common protocol (SMETA) and is uploaded onto **Sedex database**

Independent Activity: AP Member accepts supplier's corrective action plans using the SMETA guidelines (remediation timelines)

Classified - Internal use

Achievements to date

AIM® Association des Industries de Marque / European Brands Association / Europäischer Markenverband

- **Performance Improvement through mutual recognition**
  - 16 companies involved in Mutual Recognition & audit protocols benchmarked
  - List of more than 4500 audited suppliers compiled for sharing (number growing)
  - Audit recognition taking place
    - Some members are getting as much as 1/3 of the audits done through mutual recognition
    - Every 100 duplicated audits avoided = €250,000 savings
- **Systems**
  - 11 Companies involved
    - Common use of single database / data warehouse
    - ~ 5000 supplier registrations in progress
- **Sector Initiatives**
  - Agricultural sector - Engagement with Bonsucro
- **Supplier Capacity Building**
  - Supplier training sessions engaged more than 1000 supplier companies globally
  - 2011 Supplier engagement will build off of 2010 platform

Classified - Internal use

Aim-Progress Forum Is Enabling

AIM®
Association des Industries de Marque
European Brands Association
Europäischer Markenverband

Aligned Vision

Aligned Objectives

Aligned Processes

Aligned Systems

Aligned Communication

Shared Work

Customer ⇔ Member ⇔ Supplier ⇒

Shared Success

Classified - Internal use

Key Messages for FMCG Suppliers

AIM®
Association des Industries de Marque
European Brands Association
Europäischer Markenverband

Participation is a requirement

Gaps are expected

You are expected to close gaps in a reasonable way and time
. . . Continuous improvement focus … not cut and run.

We need you to take ownership
… and, overtime, apply to your suppliers and back to the origin

We won't have sustainable businesses unless we are all effective at driving Respect for human rights though the four pillars of

- **Human rights and labor standards**
- **Health and safety**
- **Environmental compliance**
- **Business integrity**

Classified - Internal use

SUSTAINABILITY
CONSORTIUM

Increased Consumer Demand for Sustainable Products

SUSTAINABILITY
CONSORTIUM

- Confusion on what constitutes a sustainable product
- Proliferation of single attribute eco-labels – need for a more holistic approach
- Lack of transparency in how products are evaluated
- Increase in green claims and green washing
- Understanding what is important to the consumer regarding sustainability

WHY WE FORMED

SUSTAINABILITY CONSORTIUM

Identifying Economic & Environmental Inefficiencies in the Supply Chain

- No set standard for suppliers to **measure** the sustainability of products throughout the life cycle.
- No set standard for suppliers to **report** measurements to their customers.
- Need for a prioritization of which products to measure.
- Gaps in data for consumer-use and post-consumer use phases.

WHY WE FORMED

SUSTAINABILITY CONSORTIUM

Emerging Global Regulations

- Communicating / labeling sustainability of products
- Need for scientifically grounded and transparent data
- Lack of a harmonized approach in a world of multi-national companies
- Need to evaluate the benefits and unintended consequences of new regulations
- Incentives and policies that promote safe and responsible innovations and strategies

WHY WE FORMED

The Sustainability Consortium

SUSTAINABILITY
CONSORTIUM

Vision

Long-term desired state

To be an independent organization of diverse global participants contributing to a more sustainable world through better products, consumption, and supply chains.

Mission

The way we achieve the vision

To develop and promote science and integrated tools that improve informed decision making for product sustainability.

Values

Shared Amongst Stakeholders

- Collaboration of diverse participants
- Scientific integrity
- Comprehensiveness and holism
- Transparency and accessibility
- Progress and solutions orientation

WHO WE ARE

How our Work Benefits Everyone

The Sustainability Consortium is working towards a more sustainable future, driving improved consumer product sustainability and creating benefits for manufacturers, retailers, consumers, and the planet. Our working groups and initiatives are collaborating to build globally accepted standardized measurement and reporting.

In other words…

SUSTAINABILITY CONSORTIUM

- TSC does Three Things
  1. We Bring Stakeholders Together
  2. We Develop SMRS (Sustainability Measurement Reporting Systems)
  3. We conduct research for product Sustainability

The Sustainability Consortium

SUSTAINABILITY CONSORTIUM

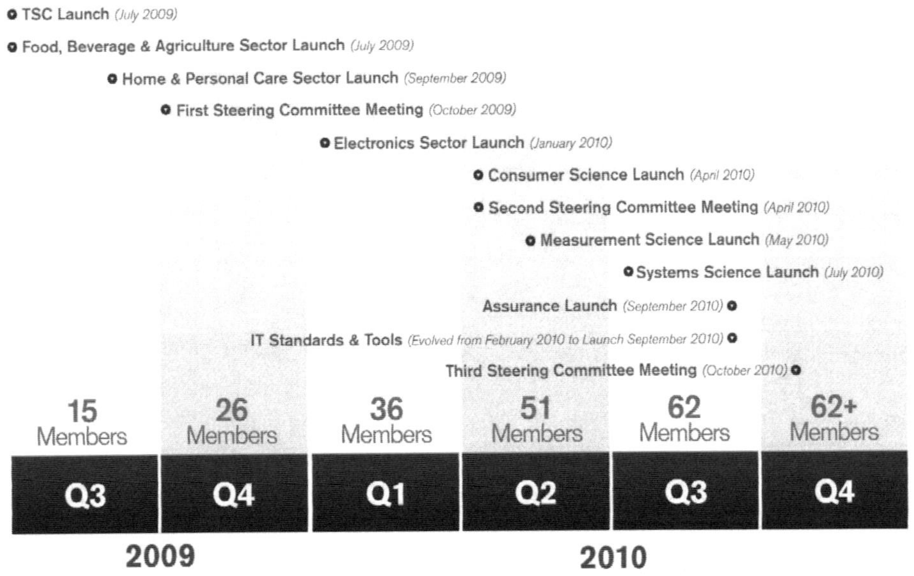

WHO WE ARE

Our Founding Members

SUSTAINABILITY CONSORTIUM

ALCOA · BSR · BASF The Chemical Company · BAYER

BEST BUY · Cargill · Church & Dwight Co., Inc. · THE CLOROX COMPANY · Coca-Cola

COLGATE-PALMOLIVE · DMI Dairy Management Inc. · DARDEN · defra · DELL

Disney · Earth Friendly Products · United States Environmental Protection Agency · General Mills · Georgia-Pacific

Henkel · hp · Kellogg's · Kimberly-Clark · KPMG

WHY WE FORMED · WHO WE ARE · WHAT WE DO · WHY IT MATTERS

Our Founding Members

SUSTAINABILITY CONSORTIUM

L'ORÉAL · MARS · MillerCoors · MONSANTO · NSF

NICE-PAK The Global Wet Wipe Experts · novozymes Rethink Tomorrow · OCTAL · PE AMERICAS · PEPSICO

P&G · SAP · SAFEWAY · SC Johnson A Family Company · SCS Scientific Certification Systems

seventh generation · Stonyfield Organic · syngenta · Tetra Pak Protects What's Good · Tyson

UL Environment · Unilever · WWF · Walmart · WM Waste Management

WHY WE FORMED · WHO WE ARE · WHAT WE DO · WHY IT MATTERS

© 2010 The Sustainability Consortium | www.sustainabilityconsortium.org

Our Tier 1 Members

SUSTAINABILITY CONSORTIUM

PRé product ecology consultants

Quantis Sustainability counts

FUNDACIONCHILE

ecoinvent

See all of our members at www.sustainabilityconsortium.org/members

Academic Partners

SUSTAINABILITY CONSORTIUM

ASU Arizona State University

University of Arkansas

Berkeley University of California

University of Cambridge

UC Davis University of California

Duke University

Harvard

MIT Massachusetts Institute of Technology

Michigan

UMass Lowell

University of New Hampshire

NC State

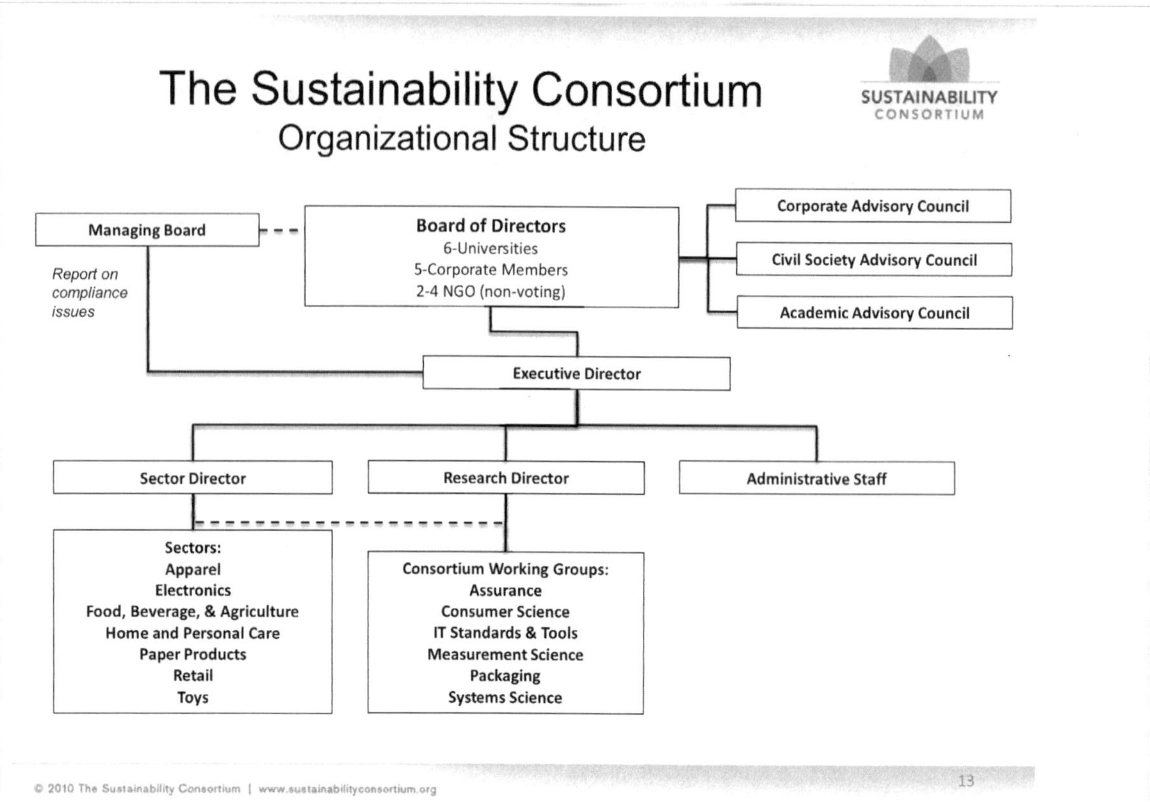
The Sustainability Consortium
Organizational Structure
SUSTAINABILITY
CONSORTIUM
Managing Board
Report on compliance issues
Board of Directors
6-Universities
5-Corporate Members
2-4 NGO (non-voting)
Corporate Advisory Council
Civil Society Advisory Council
Academic Advisory Council
Executive Director
Sector Director
Research Director
Administrative Staff
Sectors:
Apparel
Electronics
Food, Beverage, & Agriculture
Home and Personal Care
Paper Products
Retail
Toys
Consortium Working Groups:
Assurance
Consumer Science
IT Standards & Tools
Measurement Science
Packaging
Systems Science
© 2010 The Sustainability Consortium | www.sustainabilityconsortium.org
13

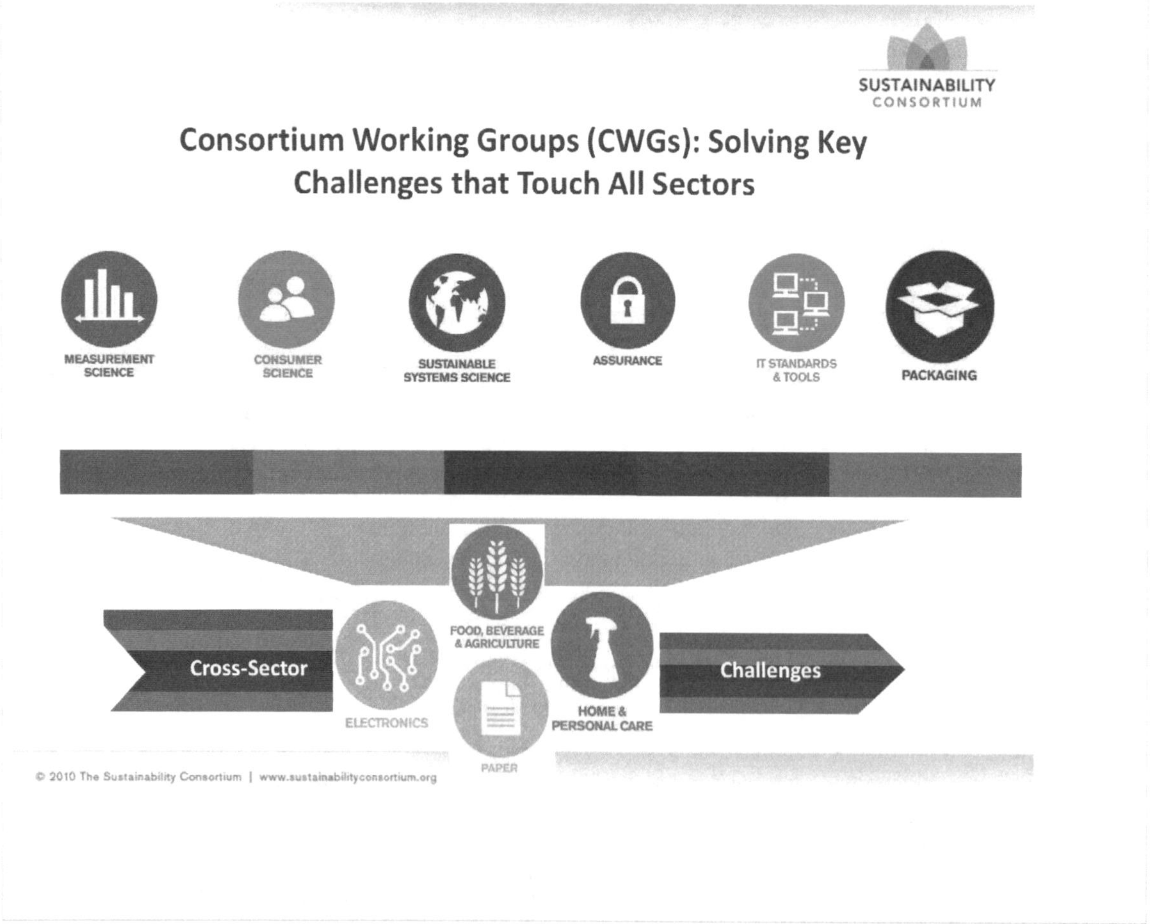
SUSTAINABILITY
CONSORTIUM
Consortium Working Groups (CWGs): Solving Key Challenges that Touch All Sectors
MEASUREMENT SCIENCE
CONSUMER SCIENCE
SUSTAINABLE SYSTEMS SCIENCE
ASSURANCE
IT STANDARDS & TOOLS
PACKAGING
Cross-Sector
Challenges
FOOD, BEVERAGE & AGRICULTURE
ELECTRONICS
HOME & PERSONAL CARE
PAPER
© 2010 The Sustainability Consortium | www.sustainabilityconsortium.org

SUSTAINABILITY CONSORTIUM

TSC Sectors

ELECTRONICS	FOOD, BEVERAGE & AGRICULTURE	HOME & PERSONAL CARE
# Members: **19**	# Members: **45**	# Members: **27**
# Participants: **25**	# Participants: **128**	# Participants: **78**
Launch Date: **Jan. 2010**	Launch Date: **Jul. 2009**	Launch Date: **Sept. 2009**
SMRS Prototype(s): **Laptop, Television**	SMRS Prototype(s): **Wheat Cereal, Flavored Yogurt, Fruit Juice**	SMRS Prototype(s): **Laundry Detergent, Shampoo, Household Cleaner**

Working Group Contact:
Carole Mars
Carole.mars@asu.edu

Working Group Contact:
Dr. Sarah E. Lewis
slewis@walton.uark.edu

Working Group Contact:
Joby Carlson
Joby.carlson@asu.edu

SUSTAINABILITY CONSORTIUM

Identifying Key Challenges

SUSTAINABILITY
CONSORTIUM

Common Challenges Face All Sectors

Selecting Methodologies and Reporting Categories

Learning to Communicate and Change Consumer Practices

Identifying Cross-Sector Commonalities

Developing for Verification Purposes

Enhancing Capability to Share Information

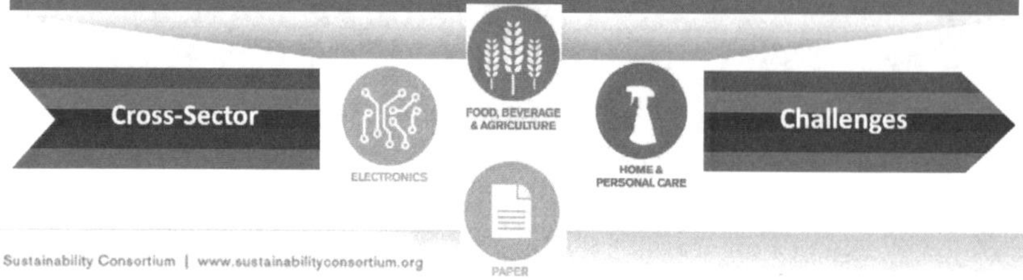

SUSTAINABILITY
CONSORTIUM

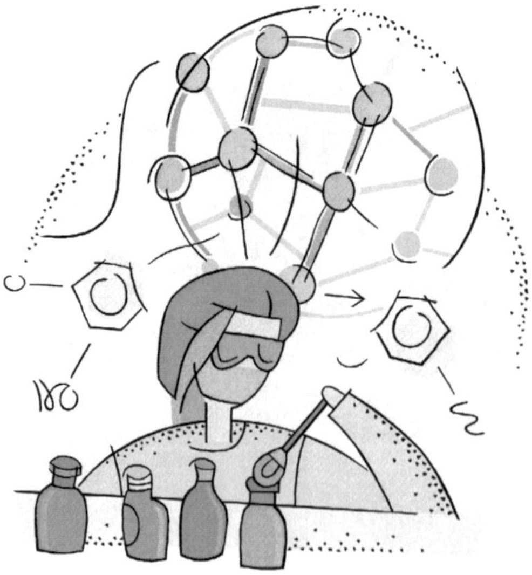

SUSTAINABILITY CONSORTIUM

Sustainability Measurement and Reporting Systems SMRS

What is an SMRS?

SUSTAINABILITY CONSORTIUM

Sustainability Measurement and Reporting System

Sustainability Product Declaration Program as defined by ISO 14025 for Type III Environmental Declarations

- Establishes a foundation that allows business to business, business to retail and business to consumer reporting.
- It addresses the questions:
  - What sustainability measures or attributes should be captured?
  - How should they be measured?
  - How should they be reported?
- Initially applied to product category or higher

More about the SMRS

SUSTAINABILITY CONSORTIUM

- Life-Cycle Assessment (LCA) based information system that facilitates decision-making about consumer goods and their supply chains.
- It is designed to facilitate the consistent exchange of sustainability related information at the product level
- It can be used by manufacturers, retailers, and eventually consumers.

3 Components to SMRS

SUSTAINABILITY CONSORTIUM

- A Product Category Rule (PCR) Registry compatible with ISO 14025
- Life Cycle Inventory (LCI) Databases linked to the PCR Registry
- Computational tools and IT infrastructure to support data sharing and reporting.

© 2010 The Sustainability Consortium | www.sustainabilityconsortium.org

Benefits to using an SMRS

SUSTAINABILITY CONSORTIUM

- Consistent framework is established that allows for a common ruler to be used for measurement of product sustinability
  - Compare products to market average
  - Prioritize risks and opportunities
  - Optimize resource allocation
- SMRS provides a product footprint and can allow you to evaluate decision alternatives based on the triggers that cause impacts

23

How is an SMRS developed?

SUSTAINABILITY CONSORTIUM

- TSC Sectors develop and validate SMRS
- TSC Sector Coordinator and stakeholders: industry members, academics, NGOs and Government agencies.
- ISO 14020 (Environmental labels and declarations — General principles)
- Valid, verified, scientifically-based, transparent, life-cycle based, innovation focused, multi-stakeholder developed, accessible
- Federated approach

WHAT WE DO

Packaging Working Group

SUSTAINABILITY CONSORTIUM

- Started January 2011
- First Meeting March 25, 2011 – Atlanta, GA
- 35 Plus Members

How will Packaging Fit in the SMRS?

SUSTAINABILITY CONSORTIUM

- Contribute information modules that will be integrated into the SMRS for product category's across sectors
- Ensure LCI Data matches identified indictors of other packaging initiatives out there

Summary

- **Creating a system** for measuring and reporting product sustainability (SMRS)
- **Research projects involving product sustainability**
  - **top researchers in environmental science and corporate social responsibility**
- **Interacting with businesses, NGOs, agencies, academia**
- **Interacting with premier sustainability firms**
- **Enhancing visibility in supply chain networks to identify value and sustainability performance**
- Developing **innovative tools, consistent and transparent science, and technical support**

Questions?

SUSTAINABILITY
CONSORTIUM
Thank You
© 2010 The Sustainability Consortium | www.sustainabilityconsortium.org
29

P&G Business Overview

Founded in 1837

$79B Sales

More than 300 branded products in > 180 countries

On-the-ground operations in > 80 countries

~127,000 Employees

Our Purpose, Values and Principles

P&G has been working on sustainability-related initiatives for over fifty years. In 2007, we added a focus on sustainability to the Company's statement of purpose. This is not a move we take lightly.

Our Purpose, Values and Principles are the guidepost for everything we do at P&G—how we conduct business, how we make decisions, how we partner externally, and how we engage our employees. Sustainability is a responsibility, especially for a Company with the scale and consumer reach of P&G. But sustainability is also a business opportunity which makes it important to shareholders.

Building a sustainability mindset into the rhythm of our business and with our employees helps enable the sustainability of our business results.

•P&G Purpose

•We provide branded products and services of superior quality and value that improve the lives of the world's consumers, **now and for generations to come.**

•P&G Principles

•We incorporate **sustainability** into our products, packaging and operations.

2012 Strategies

In 2007, P&G established five strategies for sustainability and set goals to be achieved by 2012.

Strategy 1: Products	**Strategy 2:** Operations	Strategy 3: Social Responsibility	Strategy 4: **Employees**	**Strategy 5:** Stakeholders
Delight the consumer with sustainable innovations that improve the environmental profile of our products.	Improve the environmental profile of P&G's own operations.	Improve children's lives through P&G's social responsibility programs.	Engage and equip all P&Gers to build Sustainability thinking and practices into their everyday work.	Shape the future by working transparently with our stakeholders to enable continued freedom to innovate in a responsible way.

Please refer to www.pg.com/sustainability for more information on each strategy.

2012 Sustainability Goals

2010 Report Card

Strategy	2012 Goal	Progress
Products	Develop and market at least \$50 billion in cumulative sales of "sustainable innovation products," which are products that have an improved environmental profile.	Cumulative sales of sustainable innovation products since July 2007: \$26.5 billion
Operations	Deliver an additional 20% reduction (per unit production) in CO_2 emissions, energy consumption, water consumption and disposed waste from P&G plants, leading to a total reduction over the decade of at least 50%.	see table below
Social Responsibility	Enable 300 million children to Live, Learn and Thrive. Provide 160 million days of disease and save 20,000 lives by delivering 4 billion liters of clean water in our Children's Safe Drinking Water program.	see table below

Operations	Since July 2007	Since July 2002
Energy Usage	-14%	-50%
CO_2 Emissions	-11%	-53%
Waste Disposal	-50%	-55%
Water Usage	-16%	-55%

Social Responsibility	Since July 2007
Live, Learn and Thrive	
Number of children reached	210 million
Children's Safe Drinking Water	
Liters of Clean Water Delivered	1.5 billion
Days if Disease Prevented	60 million
Lives Saved	7,450

www.pg.com

Sustainable Innovation Products

Products with a significantly reduced (>10%) environmental footprint versus previous or alternative products...and no trade-offs for the consumer

Focus on energy, water, transportation, packaging, material use

Not offset by meaningful increase in other indicators

Supported by good science and substantiated by data

Product Innovation

Cold Water Washing

Great Impact

If every U.S. household used cold water for laundry, the energy savings would be 70 to 90 billion kilowatt hours per year.

These savings would translate into ~ 34 million tons of carbon dioxide per year not released into the environment, which is nearly eight percent of the Kyoto target for the U.S.

Changing Consumer Behavior....

In the UK, 23% now wash in 30°C, up from 2%. Over 58,000 metric tons of carbon dioxide emissions prevented.

In the Netherlands, 52% now wash in **30**°C, up from 7%.

Operational Improvements

End-to-End Strategy

Supplier Engagement

Manufacturing

Finished Product Logistics

Operations has a three-tiered approach to sustainability

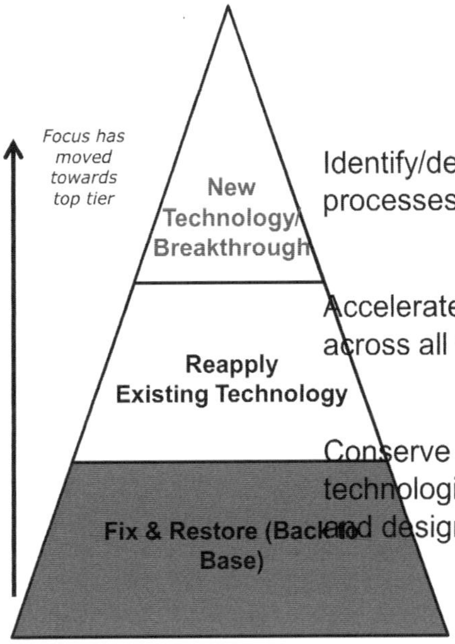

Identify/develop and implement breakthrough processes and advanced technologies

Accelerate reapplication of known opportunities across all facilities

Conserve resources by ensuring that installed technologies are operating at base conditions and design efficiency

9

Operations

Family Care Manufacturing

- **Heat Recovery Process**
  - –Recovers and re-uses heat
  - –This one project saves more energy in one year than the annual energy usage at 80% of our sites globally.
  - –Equivalent to removing the CO_2 emissions from over 2,700 cars annually*

PS Sustainability New Site Standard

The 77 point standard provide guidelines to ensure that sound environmental (sustainable) and economical choices are consistently made on all new facility construction projects

Resource Conservation Measures (RCM's) are pre-evaluated and divided into three categories:
1.Generally applicable
2.Study items
3.Not advised.

Each RCM has a worksheet which help give direction on application and benefits received.

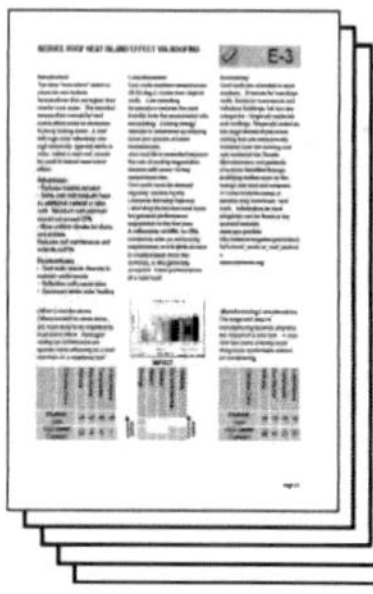

P&G's Supplier Sustainability Approach

Commercialize 360 Sustainable Innovation

- Suppliers are P&G's key source of materials, packaging, systems and ideas for sustainable innovation products.
- Suppliers bring sustainable business practice innovation—how we communicate with consumers, how we go to market and how we run our operations.
- Suppliers deliver sustainability innovations with better value and no trade-offs.
- Sustainability success requires strong collaboration across suppliers and key internal P&G customers – R&D, Marketing, Product Supply.

Measure Progress in a Simple, Consistent Way

- Use a few simple measures where P&G and suppliers can make a meaningful difference.
- Leverage existing approaches and data.
- Measures are based in science and include full life cycle analysis.
- Measures support total best value – improving sustainability without trade-offs.
- Suppliers feel rewarded for collaborating to improve P&G's sustainability (via business allocations, etc.)
- Sustainability is part of our sourcing value equation and integrated into our sourcing strategies.

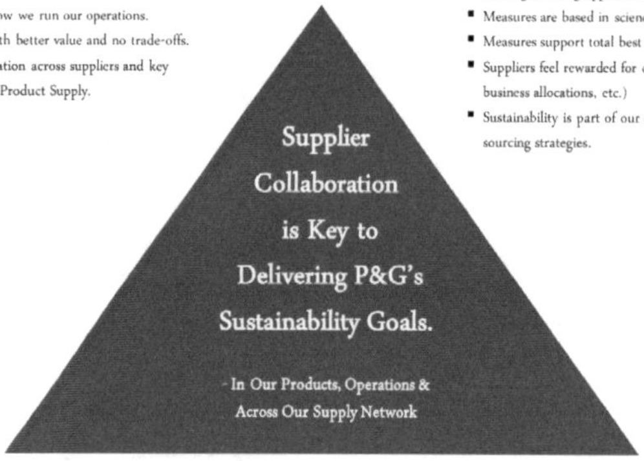

Link and Leverage to Drive Best Practices

- Benchmark with suppliers to identify best practices and reapply across the supply chain.
- Leverage our P&G Supplier Sustainability Board to guide our efforts and tap opportunities.
- Connect supplier capabilities to deliver unique sustainability improvements.
- Create a shared library of best practices and success stories.
- Partner with suppliers to influence how Consumers and Stakeholders define success.
- Proactively influence key industries to drive end to end sustainability improvements and value creation.
- Create forums to share insight, trends and approaches and ensure common understanding and approach.

Supplier Environmental Sustainability Scorecard

Why is P&G launching the scorecard?

P&G has a strong practice of improving the environmental sustainability of our products and services. The growing expectation of our consumers, customers and stakeholders is a further catalyst to extend this practice throughout our supply chain. As a first step, P&G experts have collaborated with commercial and technical resources from over twenty leading global suppliers representing broad industries, regions, and P&G spend categories (direct and indirect) to develop a new Supplier Environmental Sustainability Scorecard and rating process. Together, we have created a tool to encourage each of our suppliers to measure and improve key environmental sustainability measures.

Why doesn't P&G use a third-party like Carbon Disclosure Project (CDP) to collect this information?

We recognize and support the CDP as an important asset in encouraging the reporting of carbon-related measures. However, the CDP report does not address all scorecard areas; nor does it provide a year-on-year progress report or detail initiative support. P&G will continue to participate in CDP and encourage suppliers to do the same. Accordingly, we matched scorecard definitions and measures, wherever possible, with the same measures in CDP and Global Reporting Initiative (GRI), so suppliers will not have to do double work to track and calculate the data.

Scorecard Use

1. **Annual Supplier Sustainability Performance Ratings** (Primary Use)

   The scorecard is primarily intended to be used by P&G Buyers to determine their supplier's sustainability rating in P&G's annual supplier performance measurement process. It will be deployed in May 2010 to P&G's top suppliers. Suppliers are asked to complete the scorecard annually so they can be appropriately evaluated and rewarded for their environmental sustainability improvements and support of P&G initiatives.

2. **Supplier Business Award Decisions**

   Buyers may use components of the scorecard as a template to gather comparable sustainability measures to use in best total value business award decisions (once the scorecard is fully deployed).

   Notes:

   The scorecard template may also be used for other ad hoc activities like: tracking supplier progress against specific goals set by P&G, gathering data for studies/activities to improve the sustainability of P&G products and processes, or calculating the environmental impact of a particular supply chain at P&G.

   Suppliers are also encouraged to use the scorecard with their other customers and suppliers to further improve total supply chain environmental sustainability.

Scorecard Design

The scorecard is a concise tool for tracking a supplier's improvement on key environmental sustainability measures P&G believes are important to its supply chain. It contains the following characteristics:

- The scorecard separates "Core" (required) and "Optional" (test) data so that the measures can be tailored to suit P&G's business needs over time. P&G Spend Categories can also add other scorecard measures to capture data important to specific industries in the evaluation process.

- The definitions of the measures are aligned with worldwide standards (e.g., The Greenhouse Gas Protocol developed by the World Resources Institute and the World Business Council for Sustainable Development), where applicable.

- Flexibilities have also been built into the scorecard to account for differences in supplier's reporting capabilities, experience in environmental sustainability, size and geography. For example, we encourage suppliers to report on P&G production/service-related sustainability but accept their site-level or corporate level measures if that is the best they can do.

- Suppliers self-assess/describe their support of P&G initiatives to enable a collaborative improvement discussion with the SRO.

Supplier Environmental Sustainability Scorecard

The launch of the Supplier Environmental Sustainability Scorecard represents the next step in P&G's commitment to environmental sustainability and reflects the Company's holistic, end-to-end supply chain strategy.

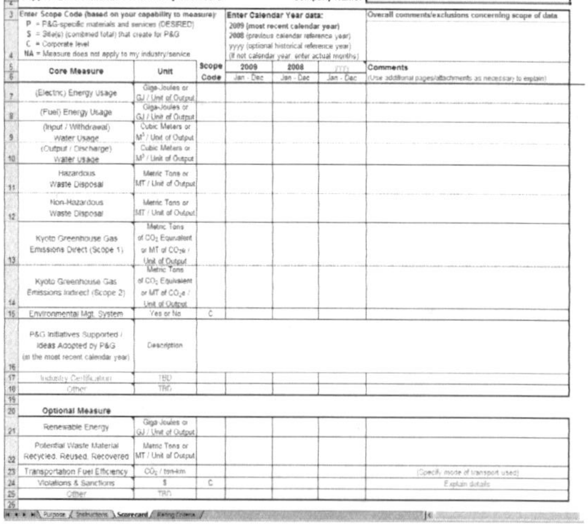

Supplier Environmental Sustainability Scorecard Company Name:

Enter Scope Code (based on your capability to measure)
P = P&G-specific materials and services (DESIRED)
S = Site(s) (combined total) that create for P&G
C = Corporate level
NA = Measure does not apply to my industry/service

Enter Calendar Year data:
2009 (most recent calendar year)
2008 (previous calendar reference year)
yyyy (optional historical reference year)
(If not calendar year, enter actual months)

Overall comments/exclusions concerning scope of data

Core Measure	Unit	Scope Code	2009 Jan - Dec	2008 Jan - Dec	YTD Jan - Dec	Comments (Use additional pages/attachments as necessary to explain)
(Electric) Energy Usage	Giga-Joules or GJ / Unit of Output					
(Fuel) Energy Usage	Giga-Joules or GJ / Unit of Output					
(Input / Withdrawal) Water Usage	Cubic Meters or M^3 / Unit of Output					
(Output / Discharge) Water Usage	Cubic Meters or M^3 / Unit of Output					
Hazardous Waste Disposal	Metric Tons or MT / Unit of Output					
Non-Hazardous Waste Disposal	Metric Tons or MT / Unit of Output					
Kyoto Greenhouse Gas Emissions Direct (Scope 1)	Metric Tons of CO_2 Equivalent or MT of CO_2e / Unit of Output					
Kyoto Greenhouse Gas Emissions Indirect (Scope 2)	Metric Tons of CO_2 Equivalent or MT of CO_2e / Unit of Output					
Environmental Mgt. System	Yes or No	C				
P&G Initiatives Supported / Ideas Adopted by P&G (in the most recent calendar year)	Description					
Industry Certification	TBD					
Other	TBD					
Optional Measure						
Renewable Energy	Giga-Joules or GJ / Unit of Output					
Potential Waste Material Recycled, Reused, Recovered	Metric Tons or MT / Unit of Output					
Transportation Fuel Efficiency	CO_2 / ton-km					(Specify mode of transport used)
Violations & Sanctions	$	C				Explain details
Other	TBD					

Purpose / Instructions / Scorecard / Rating Criteria

- Launched in May 2010 to P&G's top 400 suppliers

- The scorecard is the result of 18 months of work and close collaboration with our Supplier Sustainability Board, which includes more than 20 leading supplier representatives from our global supply chain

- The scorecard is "open code" for all to use...hopefully leading to an industry standard approach

The *Scorecard* Tab

The following diagram illustrates how a supplier should complete the scorecard for **annual supplier sustainability performance ratings**:

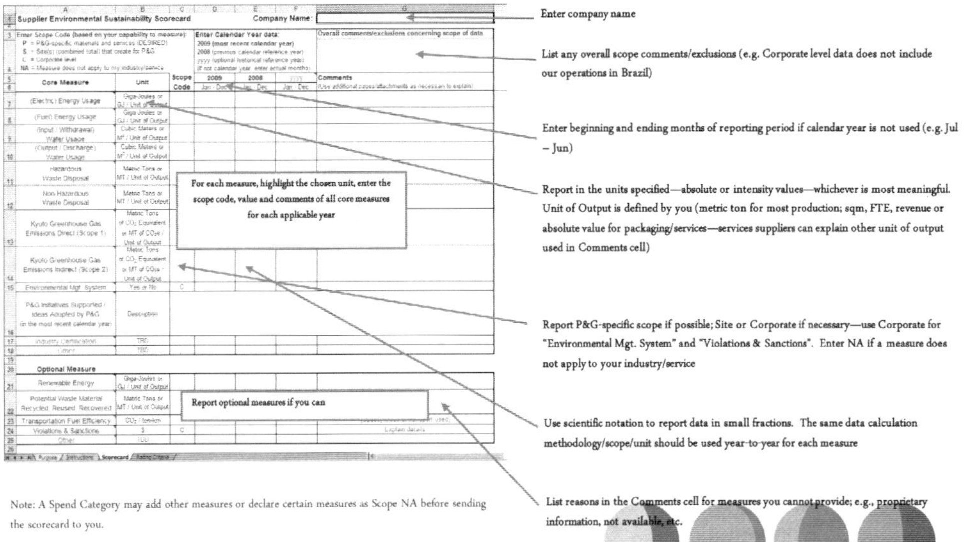

Note: A Spend Category may add other measures or declare certain measures as Scope NA before sending the scorecard to you.

The *Scorecard* Tab

Regardless of use, the desired trend for each measure is listed below:

Measure	Unit	Desired Trend (from previous year to current year)
(Electric) Energy Usage	Giga-Joules or GJ / Unit of Output	↓ Down
(Fuel) Energy Usage	Giga-Joules or GJ / Unit of Output	↓ Down
(Input / Withdrawal) Water Usage	Cubic Meters or M3 / Unit of Output	↓ Down
(Output / Discharge) Water Usage	Cubic Meters or M3 / Unit of Output	↓ Down
Hazardous Waste Disposal	Metric Tons or MT / Unit of Output	↓ Down
Non-Hazardous Waste Disposal	Metric Tons or MT / Unit of Output	↓ Down
Kyoto Greenhouse Gas Emissions Direct (Scope 1)	Metric Tons of CO2 Equivalent or MT of CO2e / Unit of Output	↓ Down
Kyoto Greenhouse Gas Emissions Indirect (Scope 2)	Metric Tons of CO2 Equivalent or MT of CO2e / Unit of Output	↓ Down
Environmental Mgt. System	Yes or No	Yes
Renewable Energy	Giga-Joules or GJ / Unit of Output	↑ Up
Potential Waste Material Recycled, Reused, Recovered	Metric Tons or MT / Unit of Output	↑ Up
Transportation Fuel Efficiency	CO2 / ton-km	↓ Down
Violations & Sanctions	$	↓ Down

Supplier Sustainability Rating Criteria

P&G's Annual Supplier Sustainability Performance Rating Criteria

Q: Does the supplier report on and improve P&G's environmental sustainability core measures, and support P&G initiatives?

Rating	5 Far Exceeds Expectations	4 Exceeds Expectations	3 Meets Expectations	2 Below Expectations	1 Far Below Expectations
Rating Criteria	All P&G scorecard core measures are reported*	All P&G scorecard core measures are reported*	Most P&G scorecard core measures are reported*	Some P&G scorecard core measures are reported*	No P&G scorecard core measures are reported*
	All P&G scorecard core measures are improved*	Most P&G scorecard core measures are improved*	Some P&G scorecard core measures are improved*	No P&G scorecard core measures are improved*	
	Worked collaboratively on all applicable P&G initiatives	Self-initiated support of all applicable P&G initiatives	Supported all initiatives requested by P&G	Supported some initiatives requested by P&G	Did not support any P&G initiatives
	Jointly developed several ideas that were adopted by P&G	Brought at least one idea that was adopted by P&G			
	Reported all additional, feasible measures requested by P&G				

*There is no adverse impact to suppliers that exclude measures that do not apply to their industry.

Purpose | Instructions | Scorecard | Rating Criteria

- Sustainability has been built in to our annual supplier rating and feedback process as a Key Performance Indicator (KPI) since 2008.
- With this scorecard, we have made the rating criteria of the Sustainability KPI more data-based, with emphasis on tracking and improving core environmental sustainability measures, and supporting P&G's sustainability initiatives.

Willie K. Johnson
Global Product Supply Sustainability
8256 Union Center Bvld.
West Chester, OH 45069
Johnson.WK@PG.COM
513-613-9879 (Office)

Sustainable Packaging

Avoiding Myopia

Randy Boeller
HP Package Engineering Program Mgr.
17th March, 2011

AGENDA

- Definition & History of Sustainability
- Elements of Sustainability
- The whole not the parts
- HP examples and lessons
- What's Next?

DEFINITION

"Sustainable development is development that meets the needs of the present without compromising the ability of future generations to meet their own needs."

–World Commission on Environment and Development 1987

3 HP Confidential

CORPORATE RESPONSIBILITY

- –51 of the largest 100 economies in the world are now corporations, only 49 are nation states
- –HP does business in 178 countries, including some that are under-developed, creating leverage points some governments do not have
- –With global reach and influence comes global responsibility

4

4 HP Confidential

SUSTAINABILITY EVOLUTION

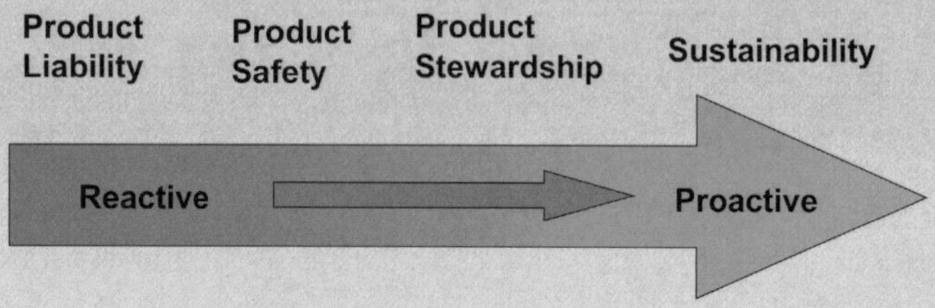

5 HP Confidential

Elements of Sustainability

6 ©2009 HP Confidential

SUSTAINABILITY COMPONENTS

WHY ENVIRONMENTAL SUSTAINABILITY MATTERS

- HP's commitment to corporate environmental responsibility started with Bill Hewlett and Dave Packard in 1939.
- It's a competitive requirement in a global market.
- Complex environmental standards and government regulations exist worldwide.
- Customers and stakeholders demand change.
- It's the right thing to do!

WALKING THE TALK

–Improving supply chain SER performance requires sustained commitment

– Work with high-risk suppliers

- Integrating SER requirements into sourcing operations
- Helping suppliers build their SER competencies
- Collaborating with NGOs, governments and industry peers
- Reporting fully and transparently the aggregated results of supplier audits, remediation efforts and training

The whole not the parts

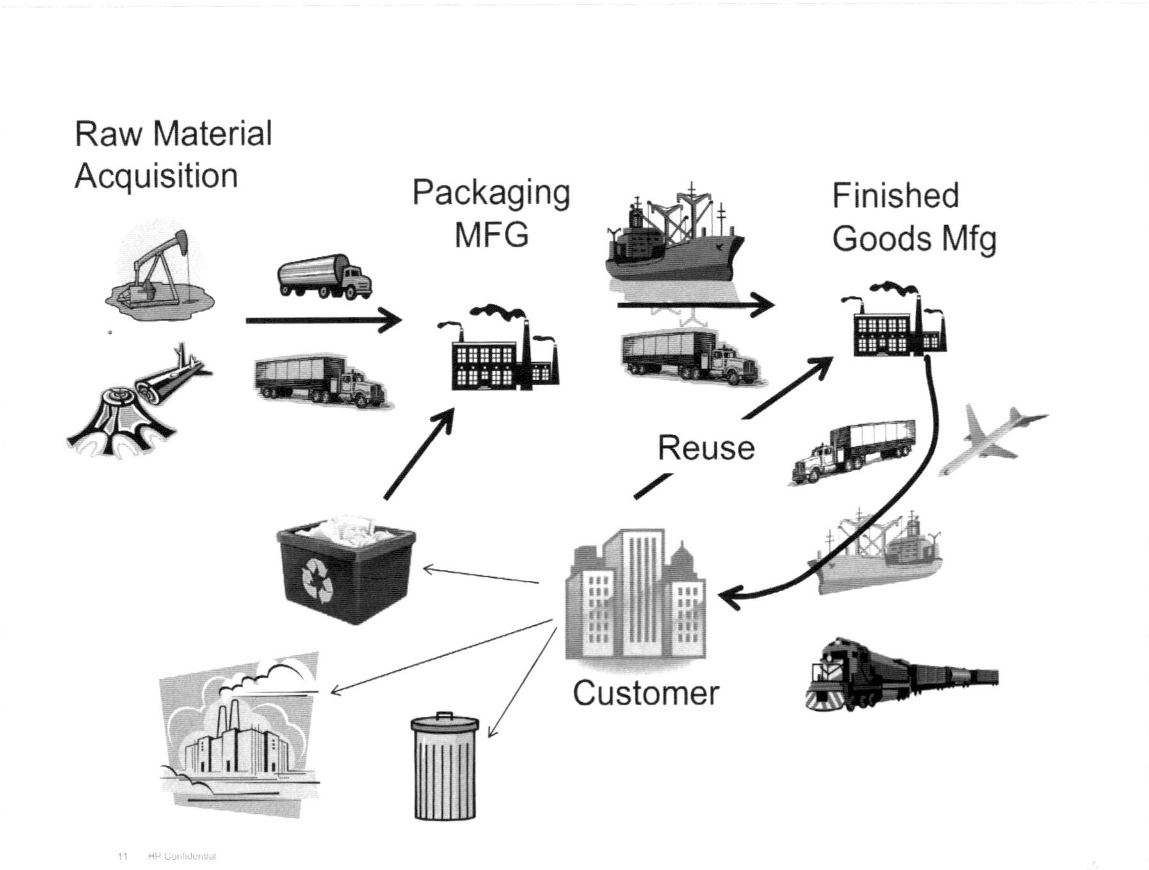
Raw Material
Acquisition
Packaging
MFG
Finished
Goods Mfg
Reuse
Customer
11 HP Confidential

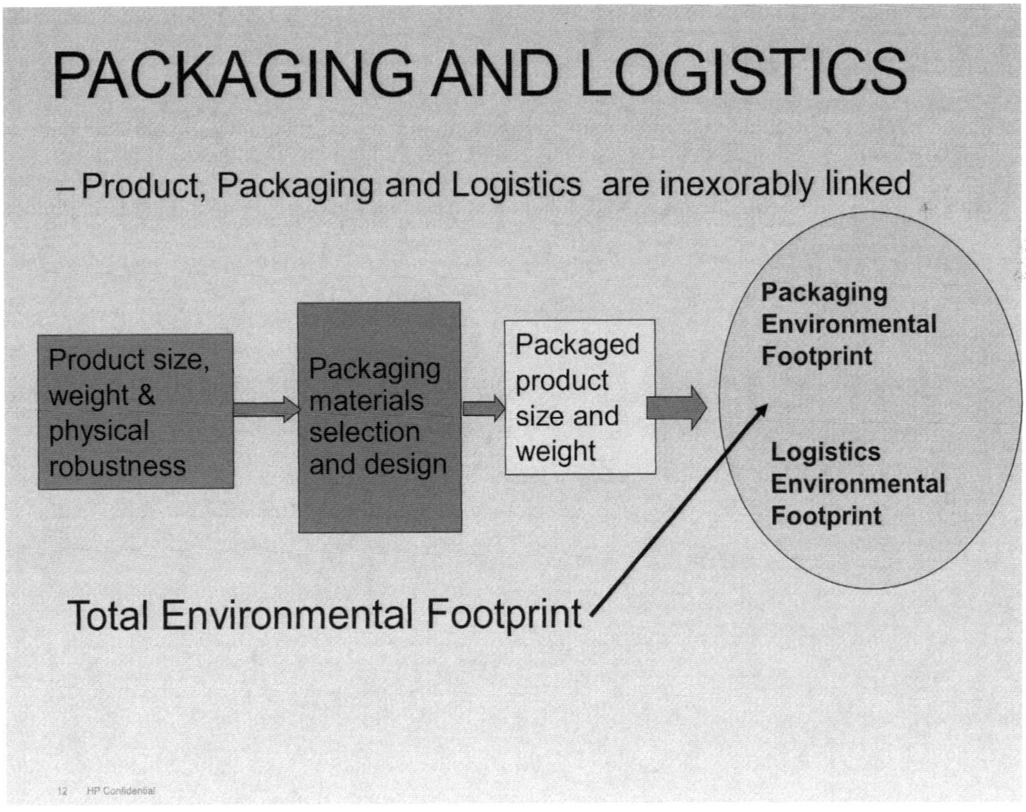
PACKAGING AND LOGISTICS
– Product, Packaging and Logistics are inexorably linked
Product size, weight & physical robustness
Packaging materials selection and design
Packaged product size and weight
Packaging Environmental Footprint
Logistics Environmental Footprint
Total Environmental Footprint
12 HP Confidential

SOME HP EXAMPLES

HP PACKAGING ENVIRONMENTAL STRATEGY

- **Remove** - Eliminate the use of regulated or environmentally problematic packaging materials.
- **Reduce** - Reduce the amount of packaging used per product.
- **Reuse** - Reuse packaging where economically and environmentally feasible while protecting the best interests of our customers and the integrity of our brand.
- **Recycle** - Take deliberate actions to increase the recycled content of packaging materials.
- **Replace** - Replace difficult to recycle packaging with more easily recycled materials.
- **Influence** - Encourage the use of sustainably sourced materials for packaging.

MATERIAL REDUCTION

–Eliminate Printed documentation

- •Paper reduction
  - – material cost savings
- •Weight reduction
  - –logistics cost reduction
- •Package size reduction and weight reduction
  - – Logistics cost reduction

15 HP Confidential

SIZE REDUCTION AND MATERIAL CHANGE

Prior solution in PVC clamshell

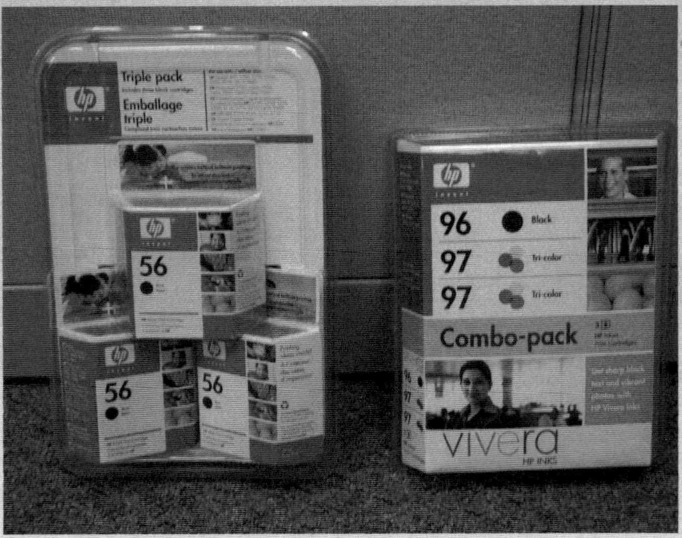

New Bulk Club clamshell in RPETG

Change frequently requires work throughout the value chain

16 HP Confidential

REDUCE AND REPLACE

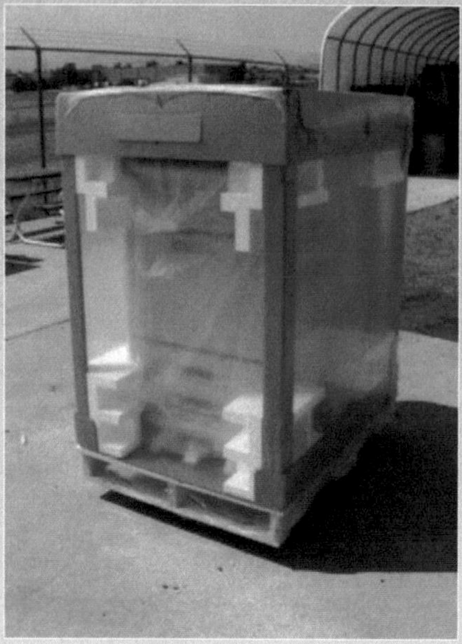

Clearview – Reduces packaging weight by 10-20 lbs per product – over 700,000 lbs in packaging eliminated in 2010.

17 HP Confidential

LIGHT WEIGHT PLASTIC PALLETS FOR AIR SHIPMENTS

Airfreight weight reduction – 2,080 Metric Tons

CO_2 reduction 120 Metric Tons

Equivalent to 2,000 cars off the road for 1 year

18 HP Confidential

PALLET RECYCLING

Polystyrene is a by-product of oil refining

Recycled High Impact →

Recycled Expanded Polystyrene is made into insulating panels used in home construction →

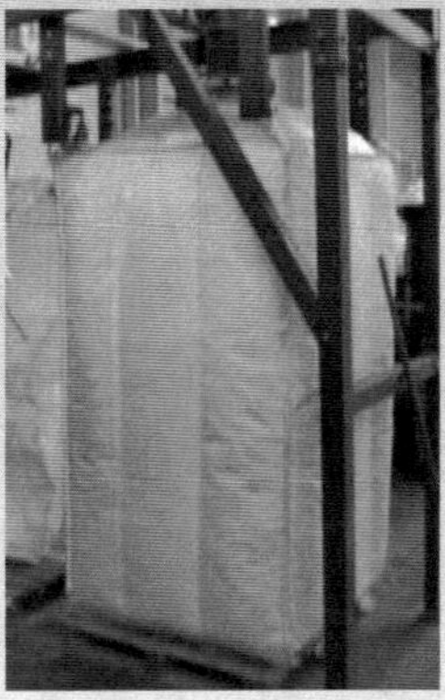

19 HP Confidential

BULK PACKAGING DESIGN

Inside the Bulk Packaging

- 8 SFF units, KBs, mice, and power cords in a single box
- 50%+ reduction in packaging material
- 50% increase in pack density

8 Units/box: 2 layers with 4 units each

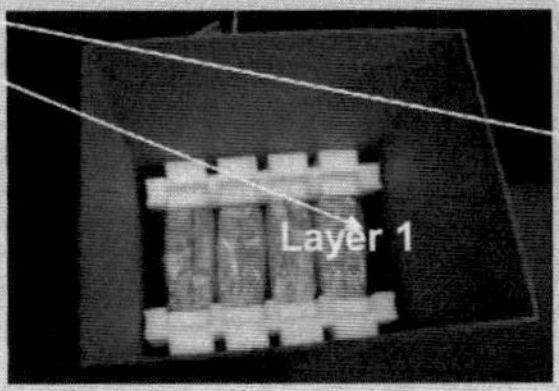

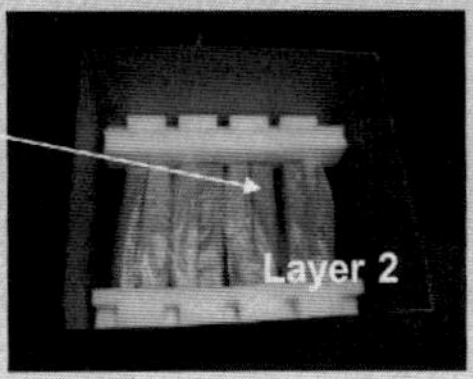

8 keyboards, mice, and power cords

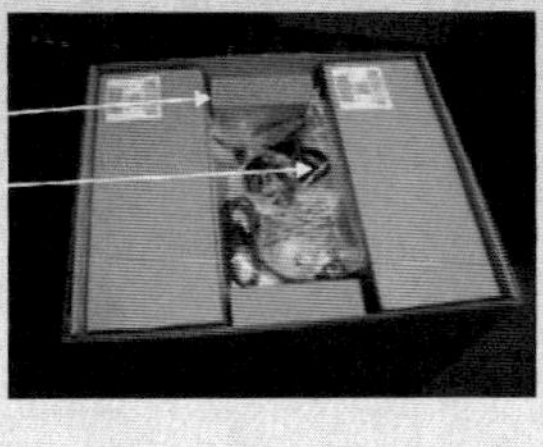

Individually Boxed Units = 24 Units

VS.

Bulk Pack Pallet = 32 Units

20 HP Confidential

What's Next

THE REGULATORY ENVIRONMENT

ENVIRONMENTAL PACKAGING REGULATION TREND

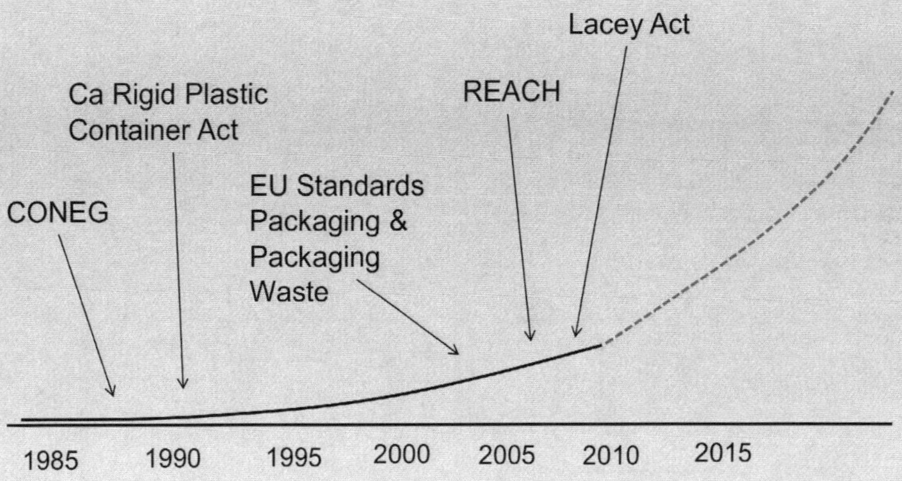

EPR

In the early 90's Germany introduced the Green Dot Packaging take-back Program

In 1994 the EU adopted Packaging and Packaging Waste Directive - 94/62/EC

Quebec, Canada - Joins Ontario as Canada's second province to establish packaging fees

Brazil, Argentina and Mexico have proposed a take-back program modeled after that of the EU

BEYOND EPR

Chain of Custody

–Lacey Act Amendment

–EU Timber Act

–Australia Timber Act

Information Collection and Reporting

Packaging materials
Packaging reduction

25 HP Confidential

If we as packaging producers and users in the US don't do more to support the building of a robust packaging recycling infrastructure local, state and federal governments will.

26 HP Confidential

THANK YOU

27 ©2009 HP Confidential

SOCIAL SUSTAINABILITY

– Fair treatment of Workers

- 700+ production suppliers in 1200+ locations
- Fair and reasonable wage
- Working Conditions

– Social Impact of manufacturing

- Conflict Minerals
- Supplier Diversity

28

28 HP Confidential

FINANCIAL RESPONSIBILITY

– Companies and corporations have a responsibility to return on the investments of owners and shareholders.

– Financial Sustainability can not be sacrificed to support Social and Environmental sustainability

– Proper Balance – Dow Jones Sustainability Index.

29

ENVIRONMENTAL SUSTAINABILITY

–Materials

–Manufacturing

–Greenhouse Gas

–Logistics

–End of Life

30

HP PLANET PARTNERS

Product take-back and recycling program

2009

- Available in more than 55 countries.
- Collected more than 3.6 million units for reuse.
- Recycled 250 million pounds of hardware and supplies.
- Surpassed 2 BILLION pounds of products recovered since 1987.
- HP earned the number 1 spot on NEWSWEEK'S Green Rankings 2009.

CO_2 AND LOGISTICS COST

CO_2 footprint = Logistics $

Other than renegotiating rates, if you have reduced your logistics cost you have reduced our carbon footprint.

The converse is true.

CARBON GOALS

–2010 Report aggregated GHG emissions from HP's largest first- and second-tier

–2012 Double voluntary purchases of renewable energy to 8 percent by 2012

–2013 HP will reduce the greenhouse gas emissions from HP-owned and HP-leased facilities by 20 percent under 2005 levels by the end of 2013 on an absolute basis.[1]

–2013 Reduce CO_2e emissions from product transport by 180,000 tones compared with 2008

LLamasoft
LLamasoft
Sustainable Packaging and Supply Chain Design – Understanding Cost and Emissions Impact of Packaging Decisions

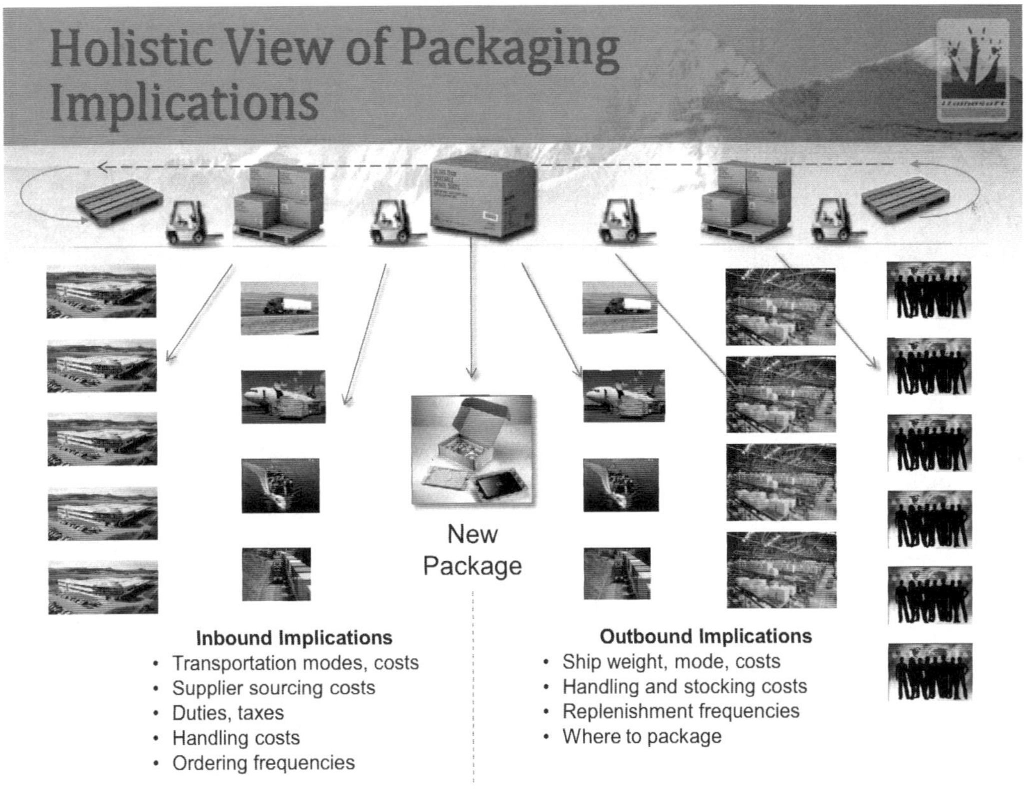
Holistic View of Packaging Implications
New Package
Inbound Implications
• Transportation modes, costs
• Supplier sourcing costs
• Duties, taxes
• Handling costs
• Ordering frequencies
Outbound Implications
• Ship weight, mode, costs
• Handling and stocking costs
• Replenishment frequencies
• Where to package

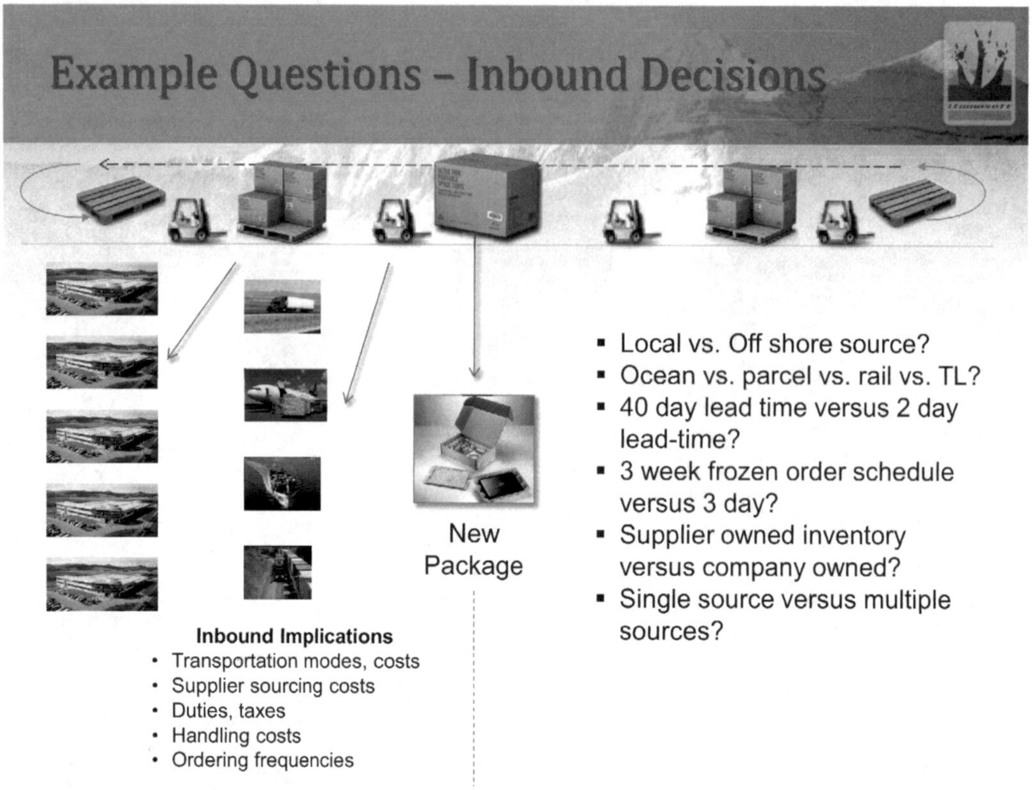
Example Questions – Inbound Decisions
New Package
Inbound Implications
• Transportation modes, costs
• Supplier sourcing costs
• Duties, taxes
• Handling costs
• Ordering frequencies
▪ Local vs. Off shore source?
▪ Ocean vs. parcel vs. rail vs. TL?
▪ 40 day lead time versus 2 day lead-time?
▪ 3 week frozen order schedule versus 3 day?
▪ Supplier owned inventory versus company owned?
▪ Single source versus multiple sources?

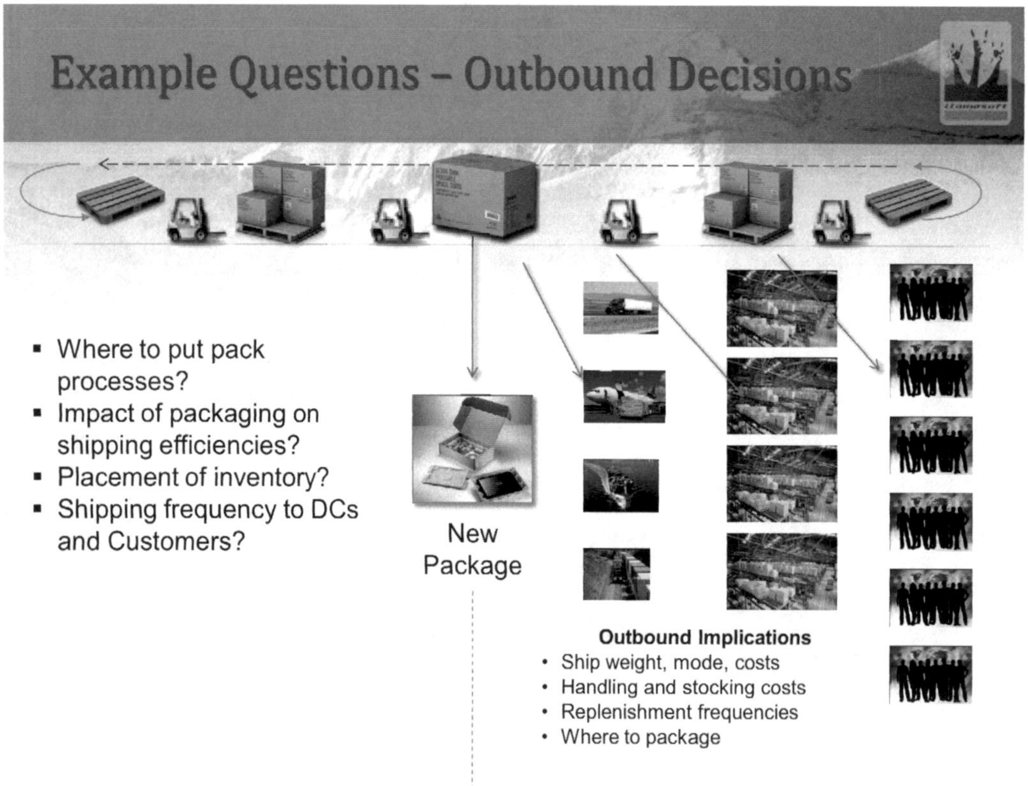
Example Questions – Outbound Decisions
▪ Where to put pack processes?
▪ Impact of packaging on shipping efficiencies?
▪ Placement of inventory?
▪ Shipping frequency to DCs and Customers?
New Package
Outbound Implications
• Ship weight, mode, costs
• Handling and stocking costs
• Replenishment frequencies
• Where to package

Understanding Total Supply Chain Costs

- Understanding trade-offs among costs
- Minimizing costs while maintaining service
- Understanding cost to serve
- For packaging and other processes

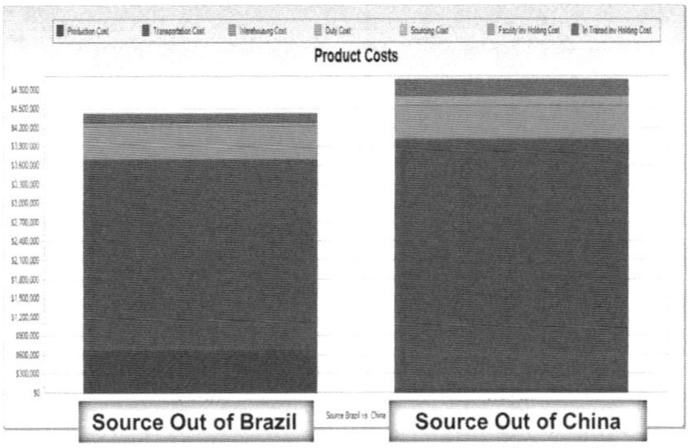

Cost to Serve – Focus on Profit as well as Costs

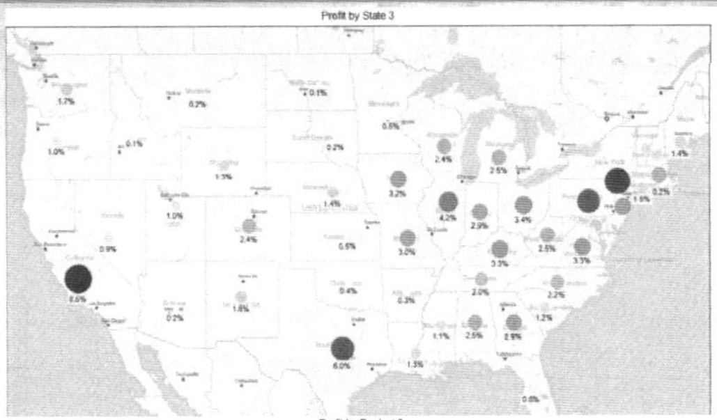

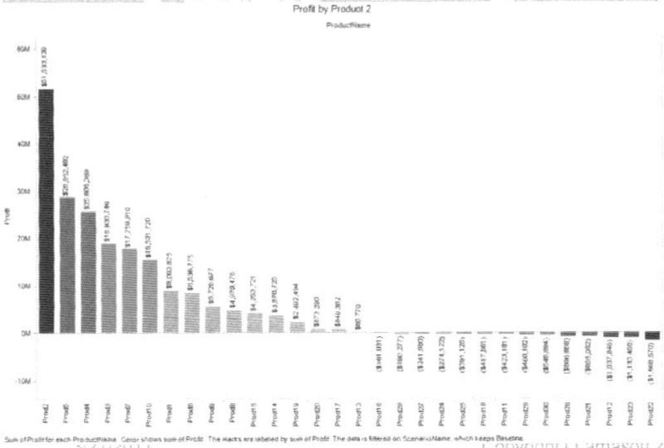

- Understand all costs associated with delivering product to customers
- Understand profitability of products and customers
- Understand optimal portfolio of products and customers

Millions in Operational Cost Savings

Global Electronics Component Manufacturer

$15 million *in supply chain cost reductions through an examination of manufacturing & distribution footprint and optimizing the product flow-paths*

Global Food & Beverage Producer

$64 million *total cost reduction in by determining the optimal location to produce and flow each product, reducing transportation costs and minimizing inter-facility moves*

Global Home Goods Store & Web Retailer

$23 million *in end-to-end cost reductions through smarter channel flow strategies for key product categories and optimized stocking locations for each product*

North American Building Supply Manufacturer

$10 million *reduction in operational costs through the rationalization of acquired assets and adjusted product distribution flow*

7

Emissions Impact from Packaging

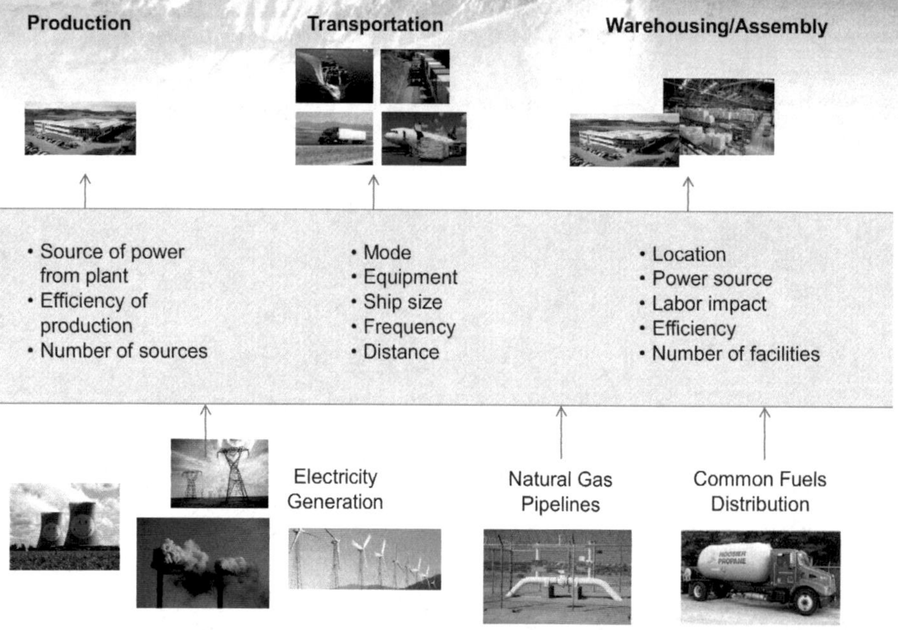

Modeling Emissions Across the Supply Chain

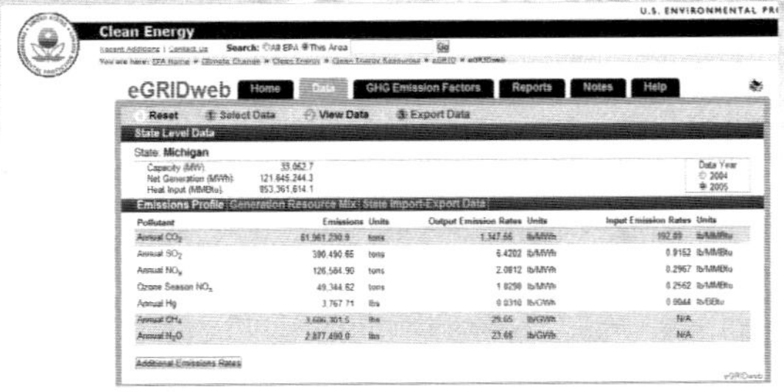

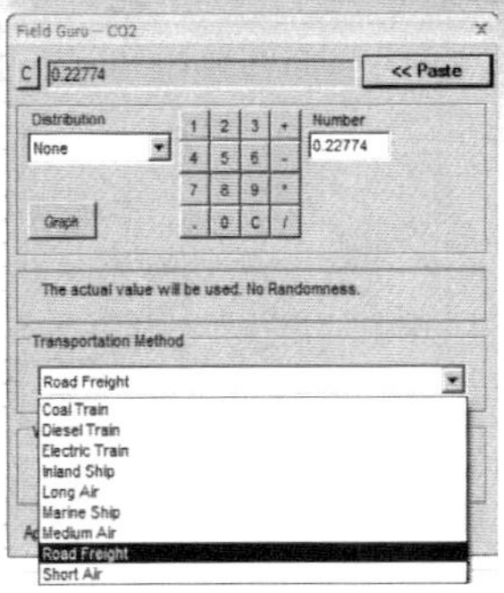

- Benchmark emissions by mode
- Benchmark emissions by source of power
- Understand carbon footprint of network design
- Constrain supply chain by emissions
- Understand tradeoffs of sourcing, transportation, and warehousing decisions

The Greenhouse Gas Protocol Initiative

The foundation for sound and sustainable climate strategies

http://www.ghgprotocol.org/

http://cfpub.epa.gov/egridweb/

Use Case Example

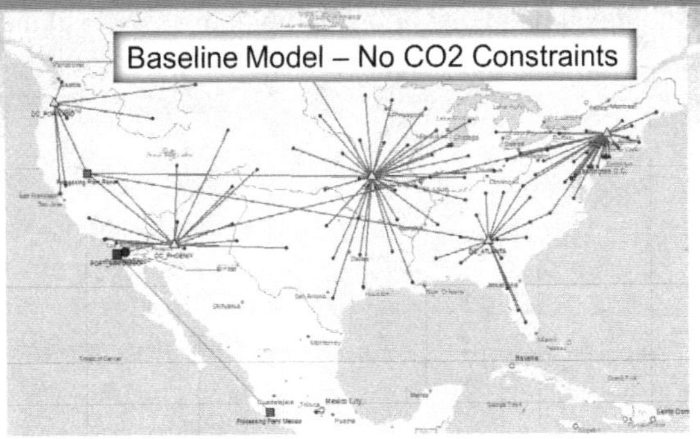

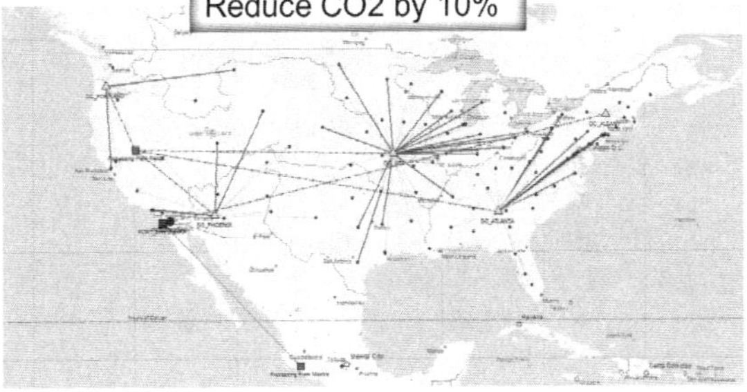

Baseline Model

- No Green carriers
- Focus on cost minimization
- Produce at lowest cost source

Reduce CO2 by 10%

- Constrain CO2 total emissions
- Greener carriers used
- Number of DCs used is reduced
- More production at greener source plants

Optimizing with CO2 Constraints

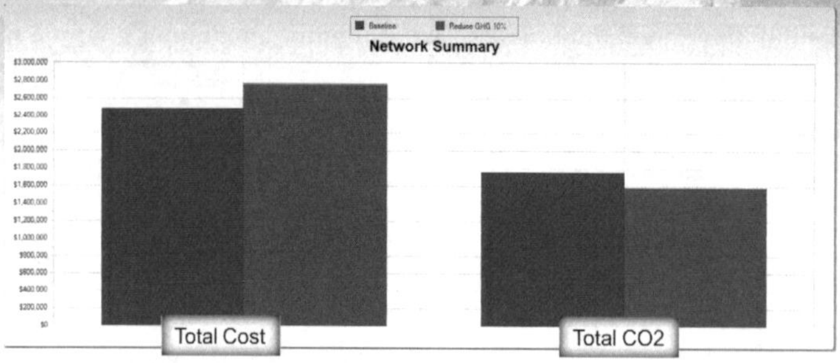

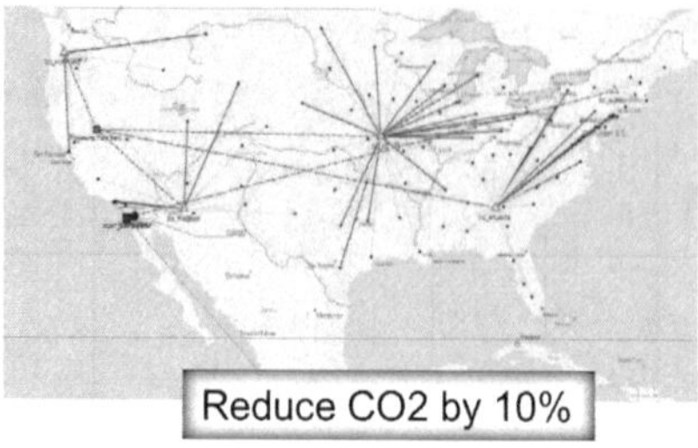

Postponement Impacts

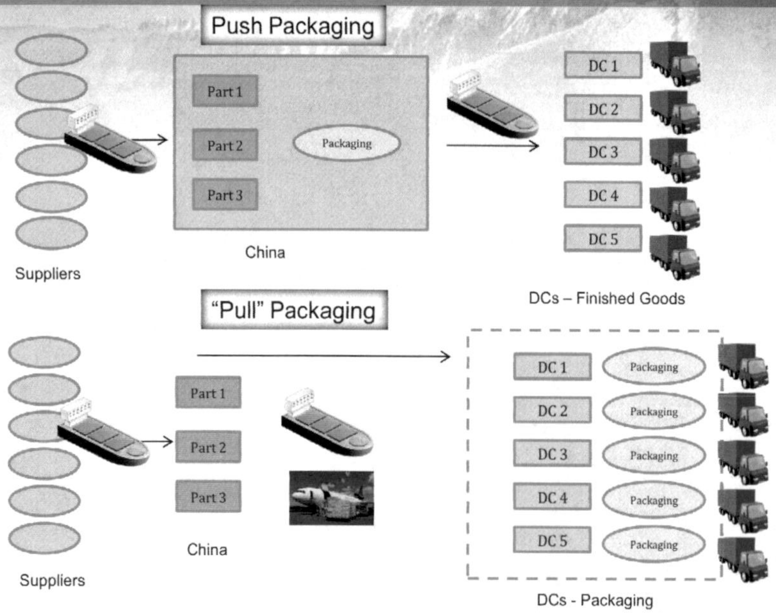

- Where to put pack processes?
- Impact of packaging on shipping efficiencies?
- Placement of inventory?
- Shipping frequency to DCs and Customers?

Analytics, Product Segmentation

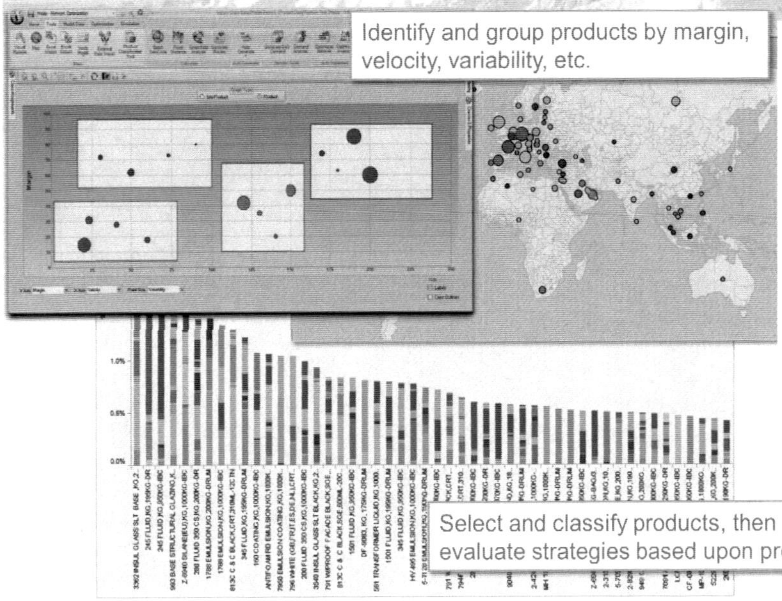

- Understand which products would be good for pull pack processes vs. push
- Identify low volume, low margin items
- Develop tailored supply chains for product segments, customers

Use Case Example

The Challenge

- Parts from China, but where should final assembly be located?
- Should a new site be started, or can existing infrastructure be used?
    - Mexico, Eastern Europe? Gut feel is China

The Approach

- Optimization used to select lowest cost/highest profit scenarios.
- Simulation used to validate results and solidify the recommendations
- Five year simulation run with changing conditions over time to test for robustness

The Outcome

- Final assembly in China meant high ship costs of FG to demand points
- Regional assembly in Mexico and Eastern Europe provided lowest total costs
- Over $40M in annual savings

Implications of Postponement

Package Close to Demand

- Higher relative labor costs
- Shorter service times
- Flexibility w/product, package, demand changes
- Lower overall transportation costs

Package in China

- Lower labor costs
- Longer service times
- Less flexibility w/product, package, demand changes
- Higher transportation costs

PlantBottle Packaging: The Next Step Forward

Vitters, S., The Coca-Cola Company

Originally presented on: 3/17/2011 14:00:00 - 14:30:00

Growth over the next ten years will not be achieved in the same way as in the past. In an era of increasing resource scarcity, environmental impact concerns and expectations from stakeholders, successful companies must find new ways to meet consumer needs while adapting to this new norm. This presentation will highlight how Coca-Cola is evolving its beverage packaging supply chain to drive long term sustainable growth by decoupling its plastic packaging from fossil-based materials and the generation of solid waste.

SUSTAINABLE PRODUCTS
+ SOLUTIONS PROGRAM
CENTER FOR RESPONSIBLE BUSINESS

University of California
Berkeley
Haas School of Business

Sustainable Products + Solutions Program
Center for Responsible Business

Sustainable Packaging Materials... What Lies Over the Horizon

Tony Kingsbury
Dow Executive in Residence,
Haas School of Business, UC Berkeley

Redefining good business.

SUSTAINABLE PRODUCTS
+ SOLUTIONS PROGRAM
CENTER FOR RESPONSIBLE BUSINESS

Sustainable Packaging Materials... What Lies Over the Horizon

Talk Outline:

- Setting the stage... for 'Over the Horizon'
- 7 Predictions for Sustainable Packaging Materials

2

SUSTAINABLE PRODUCTS
+ SOLUTIONS PROGRAM
CENTER FOR RESPONSIBLE BUSINESS

Looking Over the Horizon

Good News & Bad News

- **The Good News** – UN predicts by 2050 there will be ~9 Billion people on the planet
- This means there will be more people in need of packaged goods … food, consumer goods, etc.
- **The Bad News** – Here I refer to Thomas Friedman's book *Hot, Flat & Crowded*… These 9 Billion people will increasingly want the same stuff we have… cars, food, housing, AC, etc. etc. … and all this stuff while requiring packaging, also requires a great deal of resources to be produced.

SUSTAINABLE PRODUCTS
+ SOLUTIONS PROGRAM
CENTER FOR RESPONSIBLE BUSINESS

Another way of looking at the situation

Source: Global Footprint Network

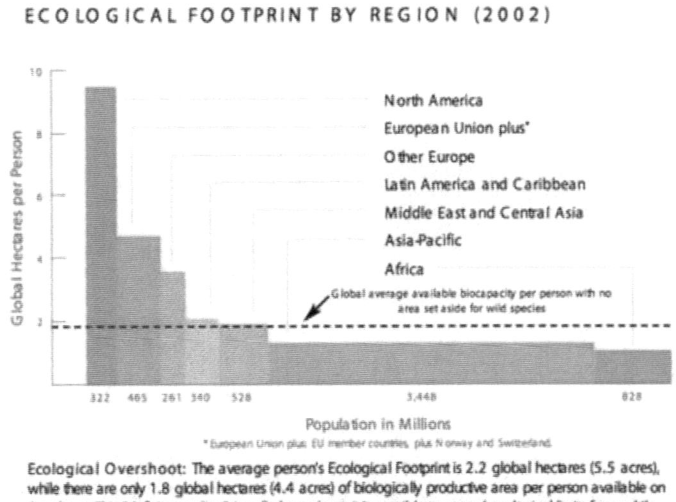

Ecological Overshoot: The average person's Ecological Footprint is 2.2 global hectares (5.5 acres), while there are only 1.8 global hectares (4.4 acres) of biologically productive area per person available on the planet. This "deficit spending" is called overshoot. It is possible to exceed ecological limits for a while, but overshoot leads to the destruction of the ecological assets on which our economy depends, resulting in depleted groundwater, collapsing fisheries, CO_2 accumulation in the atmosphere, and deforestation.

Source: Global Footprint Network

4

SUSTAINABLE PRODUCTS
+ SOLUTIONS PROGRAM
CENTER FOR RESPONSIBLE BUSINESS

We must ask ourselves

- As the world grows and becomes more affluent, will we have access to enough forests and ag waste for Pulp, Bauxite, Oil, Natural Gas, Steel, etc. to meet the needs of 9 Billion?
- I would say the answer is **NO**... especially if we continue to source and produce packaging and other goods the way we do today.
- Big Changes will be required!

5

SUSTAINABLE PRODUCTS
+ SOLUTIONS PROGRAM
CENTER FOR RESPONSIBLE BUSINESS

Over the Horizon

Economics 101

- When a material is scarce the price goes up and when prices go up either new supplies are found or consumption decreases driving to an equilibrium that balances supply and demand.
- But what happens when we struggle to find new viable sources and solutions?
- Innovation for alternatives... more efficient alternatives... thus prediction #1

6

SUSTAINABLE PRODUCTS + SOLUTIONS PROGRAM
CENTER FOR RESPONSIBLE BUSINESS

Prediction #1

- The **Most Resource Efficient Package will win**… from a system thinking standpoint
- As we move to a world of greater competition and scarcity for energy and materials … the most Carbon & Water efficient packages will be the winners in the marketplace.
- Carbon and water will drive cost and thus the most carbon and water efficient packages will win.

7

SUSTAINABLE PRODUCTS + SOLUTIONS PROGRAM
CENTER FOR RESPONSIBLE BUSINESS

Prediction #1

Carbon Efficient Package Example: Coffee

- Steel can with a LDPE #4 lid – recycled at 62%
- HDPE #2 canister – recycle 15%; WTE 15%
- Flexible Pouch - WTE 20%

2606* 3140* 1358*

* MT of CO2e per 100,000 oz. of coffee delivered

Source: ULS Report

SUSTAINABLE PRODUCTS
+ SOLUTIONS PROGRAM
CENTER FOR RESPONSIBLE BUSINESS

Prediction #1

Carbon Efficient Package Example: Coffee

- In the coffee example credit was given for lower CO2 emissions for recycling vs. virgin materials ... also credit was given for recovering the energy value of the plastics and paper. Although not as beneficial as mechanical recycling in most cases ... this does keep the molecule in play one last time.

9

SUSTAINABLE PRODUCTS
+ SOLUTIONS PROGRAM
CENTER FOR RESPONSIBLE BUSINESS

Prediction #1

Water is linked to energy and thus carbon.

- In California where I live, LBNL has estimated 20% of the energy in the State is used on WATER (to move, heat, purify, irrigate, etc.)

Example: Disposable Drink Containers vs Reusable

- Cleaning reusable glass and plastic uses lots of water and energy... Recycling is also not water and energy free.
- Equals trade-offs... where system efficiency will win

10

SUSTAINABLE PRODUCTS
+ SOLUTIONS PROGRAM
CENTER FOR RESPONSIBLE BUSINESS

Remember we are not running out of most resources

- There is as much water today as there was a thousand years ago… What we are running out of is access to clean water.
- We are not running out of aluminum and iron. They are just buried in landfills, stored in airplane shells and building structures.
- The carbon used in pulp and plastics is either buried in landfills or all around as plants or CO2 in the air. We have plenty of carbon.

11

SUSTAINABLE PRODUCTS
+ SOLUTIONS PROGRAM
CENTER FOR RESPONSIBLE BUSINESS

Prediction #2

- **Functionality will be King**
  - Even more so than today, meeting the functional needs of whatever is packaged will be more important.
  - As resources become scarce, it will be that much more important to protect the contents and not let them go to waste

12

SUSTAINABLE PRODUCTS
+ SOLUTIONS PROGRAM
CENTER FOR RESPONSIBLE BUSINESS

Prediction #3

Keeping the molecule in play will gain momentum

- What do I mean by keep the molecule in play? I mean … don't bury or litter the plastic molecules… don't bury the steel, the aluminum, the paper, etc. … keep them in play for recycling or energy recovery
  - As resources become scarce, the value of all materials will increase thus helping the economics for building out needed recycling infrastructure and participation.

13

SUSTAINABLE PRODUCTS
+ SOLUTIONS PROGRAM
CENTER FOR RESPONSIBLE BUSINESS

Recycle Content

- With more materials being recycled to keep the molecule in play…
- We will see greater use of recycled materials in packaging…
- Greater access to larger and larger piles of recycled plastic, metals, paper, etc… will lead to innovation in up-cycling these materials for use in a wide variety of packaging uses.

14

SUSTAINABLE PRODUCTS
+ SOLUTIONS PROGRAM
CENTER FOR RESPONSIBLE BUSINESS

Watch-out

- Materials that conflict with keeping the molecule in play without providing significant benefits from a resource minimization standpoint will increasing be considered unsustainable…
- Think back to my Cheese example… what if we packaged it in recyclable HDPE #2?…
- We'd need ~ ¾-1" of HDPE to maintain the needed barrier and self life

15

SUSTAINABLE PRODUCTS
+ SOLUTIONS PROGRAM
CENTER FOR RESPONSIBLE BUSINESS

Prediction #4

- Reuse will gain market share… where the system makes sense
  - This equates to a version of keeping the molecule in play.
  - Reuse keeps the box, sheet, bottle, etc. in use over and over… this is a good thing, IF…
  - …the system doesn't use more resources than alternative models
  - …the lawyers will let you take on the added liability of taking something back and cleaning it up

16

SUSTAINABLE PRODUCTS
+ SOLUTIONS PROGRAM
CENTER FOR RESPONSIBLE BUSINESS

Prediction #4

Reuse will gain market share... Example

- Why ship something in a cardboard box (that will be recycled) when the truck goes back and forth from the supplier to the retailer or distribution point? Containers that can be reused 100 times are more resource efficient.
- New Supply Chains will expand this trend. Transport to going round and round.

17

SUSTAINABLE PRODUCTS
+ SOLUTIONS PROGRAM
CENTER FOR RESPONSIBLE BUSINESS

Prediction #5

Bio-Based Packaging Materials will Grow (not necessarily biodegradable)

- This is related to carbon efficiency
- As oil and gas become more scarce and the price of carbon goes up... this will drive innovations in packaging that makes use of Bio-Sources of materials.
- We are starting to see this NOW

18

SUSTAINABLE PRODUCTS + SOLUTIONS PROGRAM
CENTER FOR RESPONSIBLE BUSINESS

Prediction #5

- Let the Bio-Bottle Wars Begin

19

SUSTAINABLE PRODUCTS + SOLUTIONS PROGRAM
CENTER FOR RESPONSIBLE BUSINESS

Prediction #5

- The bio sources we see being used will shift from food source materials to "agricultural waste materials" like bagasse, corn stover, switchgrass, algae, etc.
- We don't need the package if we don't have the food to put in it.

NatureWorks LLC

20

SUSTAINABLE PRODUCTS
+ SOLUTIONS PROGRAM
CENTER FOR RESPONSIBLE BUSINESS

Prediction #5

I predict that traditional plastics like Poly-ethylene (#2 & #4), PET, Polypropylene will dominate the bioplastics market vs. the up starts like biodegradable PLA & PHA... why?

- Cost
- Properties

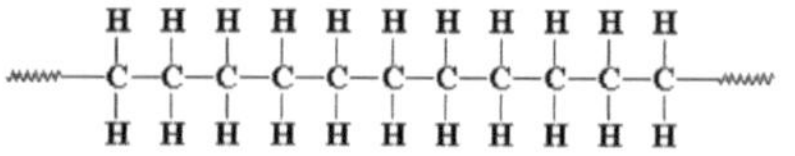

Bio = Traditional

- Efficient system (carbon efficient... not land or water efficient though)
- Composting is an inefficient way to keep the molecule in play.

21

SUSTAINABLE PRODUCTS
+ SOLUTIONS PROGRAM
CENTER FOR RESPONSIBLE BUSINESS

Prediction #6

Transparency will drive Societal full life cycle thinking about packages

- Where do materials come from? How about your energy? How about your water?

SUSTAINABLE PRODUCTS
+ SOLUTIONS PROGRAM
CENTER FOR RESPONSIBLE BUSINESS

Prediction #6

Transparency will drive Societal full life cycle thinking about packages

- How are they managed at end of life? … and who's doing it?

23

SUSTAINABLE PRODUCTS
+ SOLUTIONS PROGRAM
CENTER FOR RESPONSIBLE BUSINESS

Prediction #6

Transparency will drive Societal full life cycle thinking about packages

- How are your packages used and how do they effect the people that come in contact with them?

24

SUSTAINABLE PRODUCTS
+ SOLUTIONS PROGRAM
CENTER FOR RESPONSIBLE BUSINESS

Prediction #6

Transparency

- We now talk mainly about the base materials (plastics, paper, aluminum, etc.)... we will begin talking more about the additives, inks, dyes, glues, coatings, etc. used in packaging.
- Think about how the PVC debate shifted from the base material to phthalates additives.

25

SUSTAINABLE PRODUCTS
+ SOLUTIONS PROGRAM
CENTER FOR RESPONSIBLE BUSINESS

Prediction #6

Transparency

- **Are you ready to answer these questions about the materials you use?**

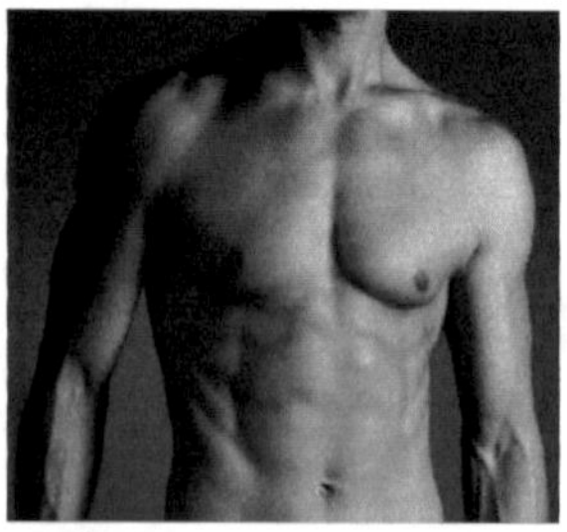

"you better get buff now,
we'll all be naked soon"

Andy Ruben, Wal-Mart

Referring to the coming trend of Transparency

26

SUSTAINABLE PRODUCTS
+ SOLUTIONS PROGRAM
CENTER FOR RESPONSIBLE BUSINESS

Prediction #7

- **Life cycle data will increasingly drive material decision making**.
  - To understand the resources
  - To understand the social implications
  - To understand the options for keeping the molecule in play…reuse, recycling, etc.
  - To deal transparency questions and decisions
  - To understand the function and system

27

SUSTAINABLE PRODUCTS
+ SOLUTIONS PROGRAM
CENTER FOR RESPONSIBLE BUSINESS

Conclusions

- Reduce… to save resources & money
- Functionality… to protect recourses
- Reuse… to keep the molecules in play
- Recycle … to keep the molecules in play
- Recover… the energy value
- BioBased will grow… but will face competition
- Transparency… know your package
- Life cycle data will drive decision making

28

SUSTAINABLE PRODUCTS
+ SOLUTIONS PROGRAM
CENTER FOR RESPONSIBLE BUSINESS

University of California
Berkeley
Haas School of Business

THANK YOU!

Tony Kingsbury
Dow Executive in Residence,
Haas School of Business, UC Berkeley
tkingsbury@haas.berkeley.edu or rakingsbury@dow.com

Redefining good business.

Sustainable Packaging: Looking Forward

Shanahan, J., Grocery Manufacturers Association

Originally presented on: 3/17/2011 15:30:00 - 16:00:00

Sustainable packaging policies are top of mind for industry and regulators and have become key drivers of innovation in our business. The consumer goods industry is working to meet demand through increasing focus on end-of-life considerations in our products and packaging as well as working to recapture much of what we produce. This session will highlight recent GMA research into sustainable waste management policy and will focus on key policy and regulatory trends affecting packaging. In addition, this session will discuss the aggregate impacts we are making toward minimizing our collective impact on the waste stream through material recovery and avoidance.

Sustainable Packaging

Does It Perform?

Larry Dull
Partner
Packaging Knowledge Group LLC
Chairman
ista Sustainability Solutions Division

Why do we want to be sustainable?

- Finite resources
  - Minerals
  - Oil
  - Water
- Corporate Triple Bottom Line
  - Financial
  - Environmental
  - Societal

What has packaging got to do with sustainability?

- Packaging utilizes valuable and limited resources
- Packaging has high visibility
  - Easy target

High Visibility of Packaging

2009 U.S. MSW

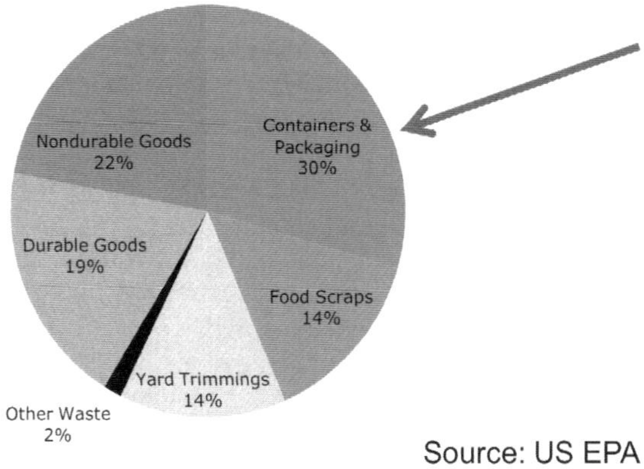

Source: US EPA

What is sustainable packaging?

- EPA suggests that packaging can be made more sustainable by:
  - Eliminating toxic constituents
  - Using less material
  - Making it reusable
  - Using more recycled content
  - Making it more readily recyclable

What is sustainable packaging?

- Sustainable Packaging Coalition suggests:
  - is beneficial, safe and healthy for individuals and communities throughout its life cycle
  - meets market criteria for performance and cost
  - is sourced, manufactured, transported and recycled using renewable energy
  - maximizes the use of renewable or recycled source materials
  - is manufactured using clean production technologies and best practices
  - is made from materials healthy in all probable end-of-life scenarios
  - is physically designed to optimize materials and energy
  - is effectively recovered and utilized in biological and/or industrial cradle-to-cradle cycles

Packaging

- Pre-industrial revolution
- Post-industrial revolution

What does package performance mean?

- Packaging is useful in today's post-industrial revolution society because it performs <u>certain functions</u>
- If packaging fails or is lacking in any of these functions, it is not performing and is therefore of <u>diminished value</u>

Packaging Functions

- Containment
- Protection
- Communication
- Utility/Safety
- Sustainability

Packaging Functions

- Containment
  - The package must contain the product to facilitate the handling of the product and its movement through the distribution system from the point of manufacture to the end-use location (and many times during storage at the end use location)

Packaging Functions

- Protection
  - The package must insure that the product gets through the tortuous distribution system and to the end-user in a usable, good quality condition. That is, the package must protect the product from the environment
  - The package must also (for some products) protect the environment from the product

Product Protection

- July 2009 survey by Genco: when shoppers see damaged packaging
  - 55% discarded the brand and 36% purchased another brand
  - Perception of the product as "the brand you trust" dropped from 73% to 41%
  - 25% of even the most brand loyal shoppers question the safety of the product

Packaging Functions

- Communication
  - The package plays an integral part in relaying information to the end-user
    - Sales messages
    - Required information (product identification, net contents, manufacturer, nutritional information, etc.)
    - Use directions
    - Reflect the value of the product
    - Recycling/reuse information

Packaging Functions

- Utility
  - The package should provide utility and/or safety in the use of the product (i.e. dispensing)
  - Provide tamper evidence
  - Provide child-proof features

Packaging Functions

- Sustainability
  - Harmonize the demands of People, Performance and the Planet
  - People (Society in TBL) – assure proper working conditions, wages, benefits
  - Planet – conserve finite resources
  - Performance – the most critical deliverable for a sustainable package

The Goal

- In today's resource-constrained environment, the goal then becomes:

  enhancing the sustainability of packaging without sacrificing any

  of the functionality

The Goal

- Put mathematically, the goal historically has been:

Product + Package = Environment

Where: Environment = Distribution Environment

The Goal

- In today's sustainability minded climate, the goal now is:

Product + Package = Environment

Where: Environment = Distribution Environment + Sustainability

The Easiest Thing To Do

- Reduce materials
  - Light weight packages
  - Switch from rigid to flexible packages
  - Downsize packages

But You Can Go Too Far

- Several life cycle analyses have looked at the environmental impacts of damaged product (i.e. from packaging that has been reduced too far in the name of sustainability)

The Bottom Line

- Sustainable packaging is good, but it can be taken too far!

ISTA

- A technical trade association, focused entirely on issues related to transport packaging and distribution
- "We help our members control costs, damage and resources during the distribution of packaged-products"

ISTA

- Sustainability Solutions Division
- Identified a general lack of resources as to how to go about designing sustainable packaging
- Core Team formed

ISTA Guide to Responsible Packaging Design

- Establish Project Scope
- Define Boundaries and Metrics
- Assess environmental impact for existing package
- Assess environmental impact for proposed package
- Select Best Package Design Option
- Prototype and Validate Proposed Design
- Assess, Recommend and Document Final Design
- Implement Final Design

ISTA Guide to Responsible Packaging Design

- Three recommended approaches:
  - Full Life Cycle Analysis
  - LCA "lite"
  - Life Cycle Thinking

Checklist

- What kind of packaging is needed (primary, secondary, tertiary, transport)?
- Does the whole system use the minimum adequate amount of material to maintain the necessary level of safety, hygiene and acceptance for the packaged product for the consumer?
- What factors limit further reductions in materials?
- Are specifications and information available for all materials?
- If part, or all, of the packaging system is intended to be re-used, is it physically capable of being re-used and is there a system in place for re-use?
- After use, will the packaging be capable of being recovered either as a material or as energy or as compost?

ISTA Guide to Responsible Packaging Design

- Time line
  - Internal consensus review by beginning of second quarter
  - External consensus review by end of second quarter
  - Document available through ista in third quarter

ISTA Guide to Responsible Packaging Design

- New ista training course
  - October 2011 – one day informational session
  - Fourth quarter 2011 – multi-day certification course
  - Watch ista website for further info

Sustainable Packaging

Does It Perform?

Yes, if it's designed "responsibly"

For Further Information

www.ista.org

larry.dull@packagingknowledgegroup.com

AMERIPEN

North American organization to advocate
for the packaging value chain with a
"material neutral" perspective

Presented by
Ron Cotterman
Executive Director, Sustainability
Sealed Air Corporation

March 2011 1

AMERIPEN Vision

AMERIPEN will become the recognized cross-sectional North American industry and trade organization to influence environmental packaging issues related to legislation and regulation.

March 2011 2

AMERIPEN Mission

AMERIPEN will **represent the North American market**, whereby best use is made of the role, benefits and functions of packaging in society, such as protection of products and the environment and supplying information to and meeting the needs of consumers and customers.

AMERIPEN, using sound science, will **lobby legislators and regulators** on issues related to the environmental impacts, marketing, sourcing, manufacture, and distribution of packaging throughout its lifecycle.

AMERIPEN will **engage and collaborate** with existing industry organizations to leverage their expertise in areas of:
- analyzing the environmental and social effects of packaging
- creating a better understanding of the role of packaging
- minimizing the environmental impact of packaging

March 2011 3

Membership

Companies which support:
- A united trade and industry organization
- Active and cooperative issue resolution
- Material and packaging system neutrality

Membership is open to all packaging value chain economic operators such as:
- Raw material producers
- Packaged goods manufacturers
- Retailer of packaged goods
- Packaging suppliers and/or converters
- Waste haulers
- Distributors of packaged goods
- Recyclers

March 2011 4

Founding Members

Company	Sector	Representative
The Coca-Cola Company	Brand Owner – Beverage	April Crow
Colgate-Palmolive Company	Brand Owner - Personal Care	Joan Pierce
ConAgra Foods	Brand Owner – Food	Gail Tavill
Dow Chemical Co.	Raw Material – Resin	Jeff Wooster
DuPont Packaging and Industrial Polymers	Raw Material – Resin	Shanna Moore
Kellogg Company	Brand Owner – Food	Kimberly Harvey
MeadWestvaco (MWV)	Raw Material /Converter	Laura Rowell
Procter & Gamble	Brand Owner – Personal Care	Joe Keller
Sealed Air Corporation	Converter – Plastics	Ron Cotterman
Tetra Pak, Inc.	Converter – Aseptic Pkg	Elisabeth Comere

March 2011 5

For further information

www.ameripen.org

March 2011 6

Opening Keynote, Day Two

Sustainable Packaging Symposium

Chicago, 16-18 March 2011

Gregory A. Norris
TSC, Harvard, New Earth

Outline

- Inside the TSC
- Looking Ahead: Within TSC and Beyond

The Current Situation

SUSTAINABILITY CONSORTIUM

- Increased consumer demand for sustainable products: B2B, B2C
- Economic & environmental inefficiencies in the supply chain
- Emerging global regulations
- Need to enhance scientific understanding of social and environmental impacts & benefits
- Inconsistent measures and reporting

WHY WE FORMED

The Sustainability Consortium

SUSTAINABILITY CONSORTIUM

Vision

Long-term desired state

To be an independent organization of diverse global participants contributing to a more sustainable world through better products, consumption, and supply chains.

Mission

The way we achieve the vision

To develop and promote science and integrated tools that improve informed decision making for product sustainability.

Values

Shared Amongst Stakeholders

·Collaboration of diverse participants
·Scientific integrity
·Comprehensiveness and holism
·Transparency and accessibility
·Progress and solutions orientation

55

The Sustainability Consortium

- TSC Launch (July 2009)
- Food, Beverage & Agriculture Sector Launch (July 2009)
- Home & Personal Care Sector Launch (September 2009)
- First Steering Committee Meeting (October 2009)
- Electronics Sector Launch (January 2010)
- Consumer Science Launch (April 2010)
- Second Steering Committee Meeting (April 2010)
- Measurement Science Launch (May 2010)
- Systems Science Launch (July 2010)
- Assurance Launch (September 2010)
- IT Standards & Tools (Evolved from February 2010 to Launch September 2010)
- Third Steering Committee Meeting (October 2010)

2009 Q3	2009 Q4	2010 Q1	2010 Q2	2010 Q3	2010 Q4
15 Members	26 Members	36 Members	51 Members	62 Members	66 Members

66

How our Work Benefits Everyone

The Sustainability Consortium is working towards a more sustainable future, driving improved consumer product sustainability and creating benefits for manufacturers, retailers, consumers, and the planet. Our working groups and initiatives are collaborating to build globally accepted standardized measurement and reporting.

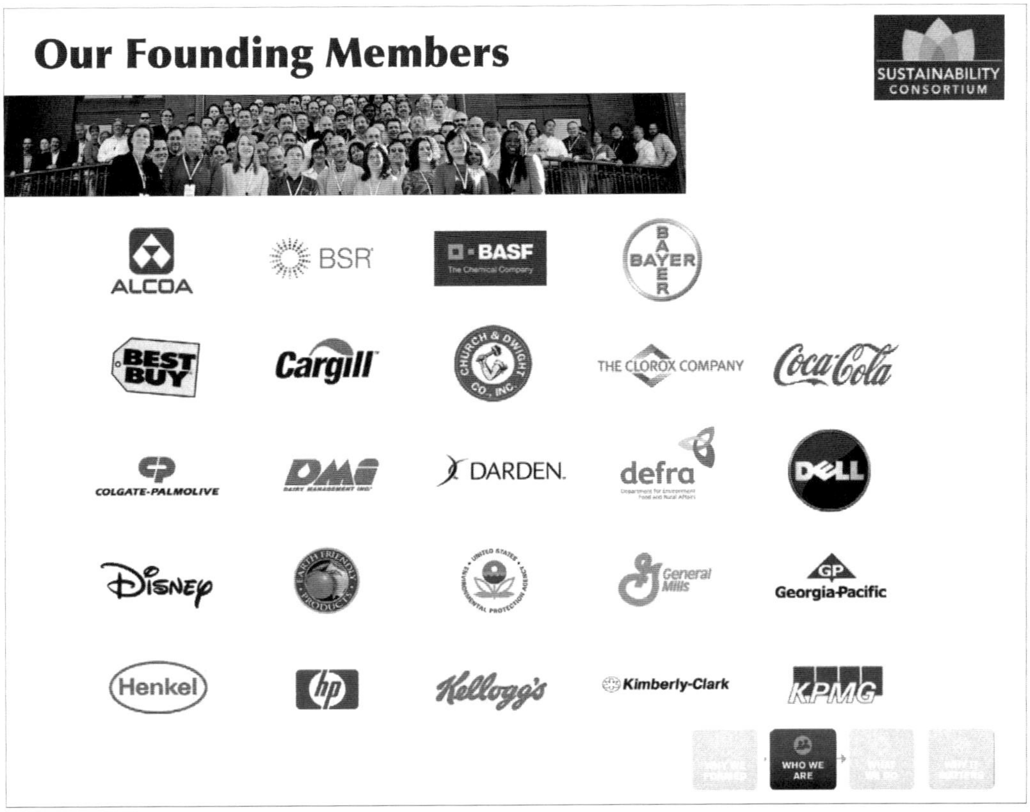
Our Founding Members
SUSTAINABILITY CONSORTIUM
ALCOA
BSR
BASF
The Chemical Company
BAYER
BEST BUY
Cargill
CHURCH & DWIGHT CO., INC.
THE CLOROX COMPANY
Coca-Cola
COLGATE-PALMOLIVE
DMI
DAIRY MANAGEMENT INC
DARDEN
defra
DELL
Disney
EARTH FRIENDLY PRODUCTS
UNITED STATES ENVIRONMENTAL PROTECTION AGENCY
General Mills
GP
Georgia-Pacific
Henkel
hp
Kellogg's
Kimberly-Clark
KPMG
WHO WE ARE

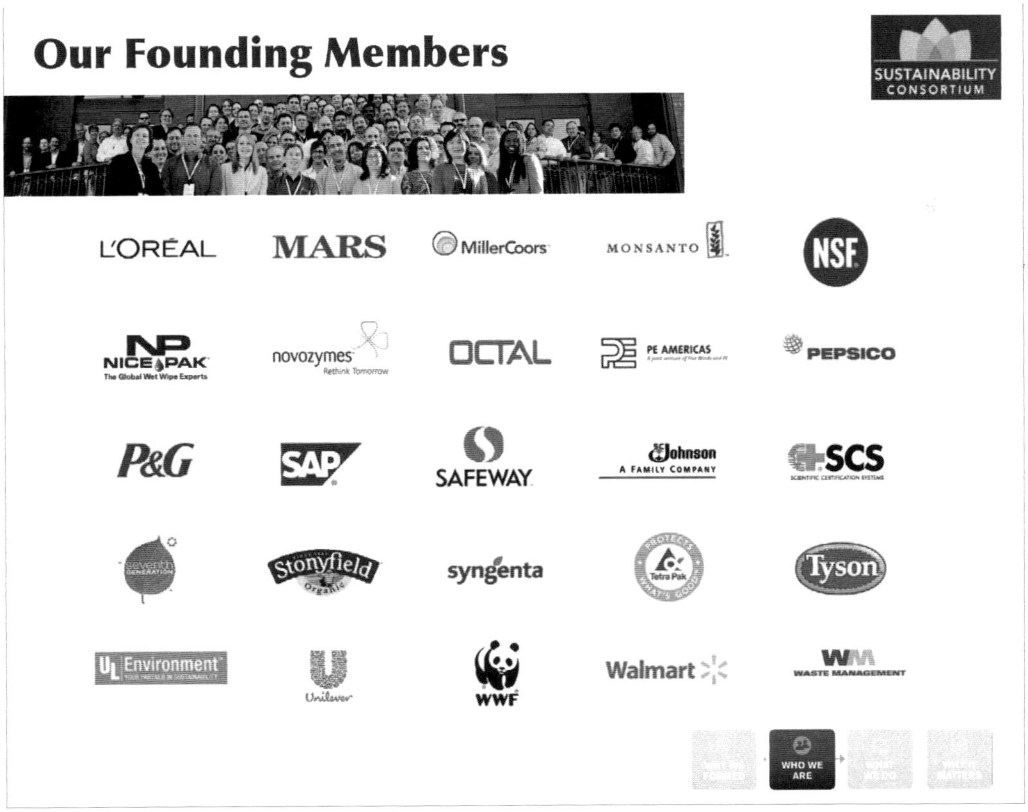
Our Founding Members
SUSTAINABILITY CONSORTIUM
L'ORÉAL
MARS
MillerCoors
MONSANTO
NSF
NP
NICE PAK
The Global Wet Wipe Experts
novozymes
Rethink Tomorrow
OCTAL
PE AMERICAS
PEPSICO
P&G
SAP
SAFEWAY
SC Johnson
A FAMILY COMPANY
SCS
SCIENTIFIC CERTIFICATION SYSTEMS
seventh generation
Stonyfield Organic
syngenta
Tetra Pak
PROTECTS WHAT'S GOOD
Tyson
UL Environment
Unilever
WWF
Walmart
WM
WASTE MANAGEMENT
WHO WE ARE

Academic Partners

SUSTAINABILITY CONSORTIUM

ASU ARIZONA STATE UNIVERSITY

UNIVERSITY OF ARKANSAS

Berkeley UNIVERSITY OF CALIFORNIA

UNIVERSITY OF CAMBRIDGE

UC DAVIS UNIVERSITY OF CALIFORNIA

Duke UNIVERSITY

VE RI TAS HARVARD

MIT Massachusetts Institute of Technology

MICHIGAN

UMASS LOWELL

UNIVERSITY of NEW HAMPSHIRE

NC S

WHY WE FORMED · WHO WE ARE · WHAT WE DO · WHY IT MATTERS

Example research projects

SUSTAINABILITY CONSORTIUM

- An overview of methods for screening chemical hazards – U Massachusetts Lowell
- Product attribute impact assessment for electronics – MIT
- Consumer knowledge and behavior about sustainable products – Harrison Group
- Agricultural land-use impacts to biodiversity – U Arkansas
- Analysis and catalogue of water resource Indices – U Michigan
-
- Assist New Earth in development of Social Hot Spot Database - UNH

Sustainability Measurement and Reporting System

What is SMRS?

Sustainability Measurement and Reporting System

Sustainability Product Declaration Program as defined by ISO 14025 for Type III Environmental Declarations

Establishes a foundation that allows business to business, business to retail and business to consumer reporting.

It addresses the questions:

- What sustainability measures or attributes should be captured?
- How should they be measured?
- How should they be reported?

Initially applied to product category or higher

How is SMRS developed?

- TSC Sectors develop and validate SMRS
  - Product Categories Rules
  - Baseline Model
  - Computational tools

• TSC Sector Coordinator and stakeholders: industry members, academics, NGOs and Government agencies.

• ISO 14020 (Environmental labels and declarations)
Valid, verified, scientifically-based, transparent, life-cycle based, innovation focused, multi-stakeholder developed, accessible

·Federated approach

Three Step SMRS Reporting

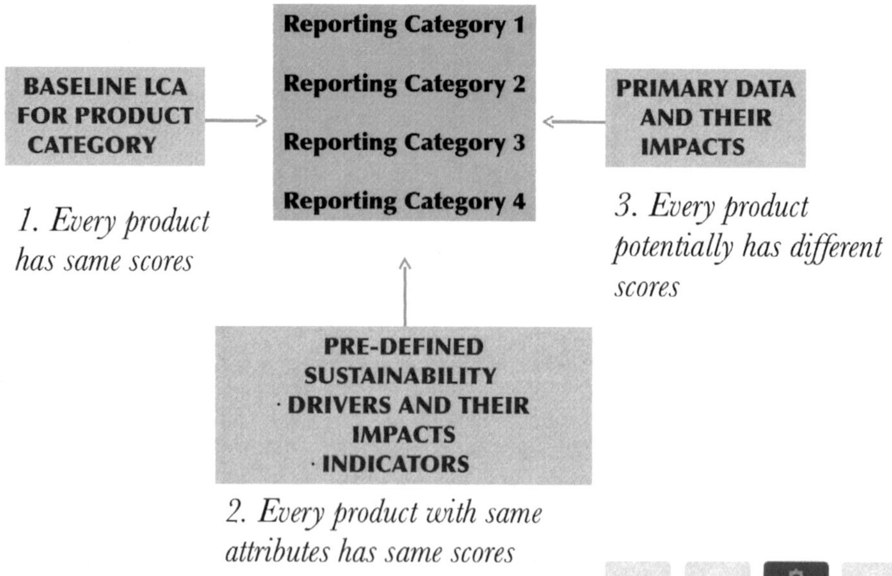

Example: Laundry detergent

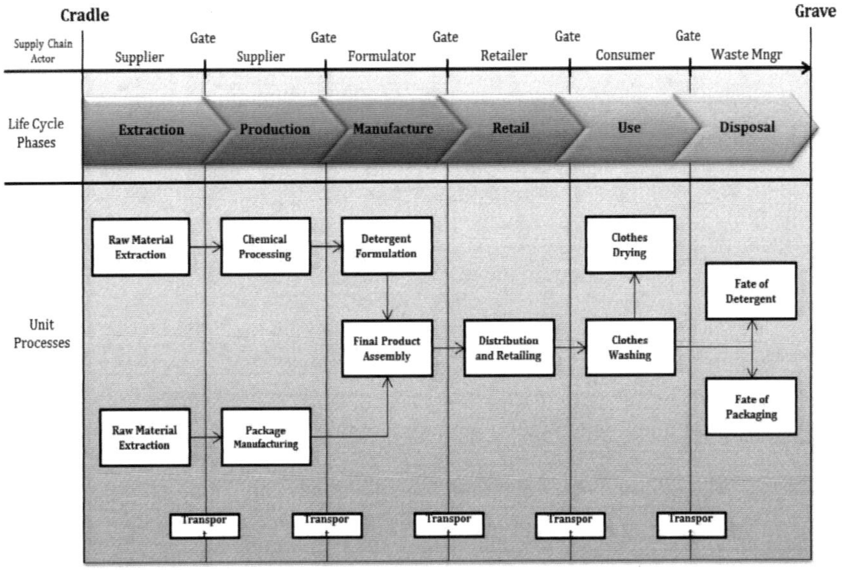

home & personal care

working group

Baseline Models: **1**
Hotspots Identified: **6**
SPDs Identified: **16**
Example: **Laundry Detergent**

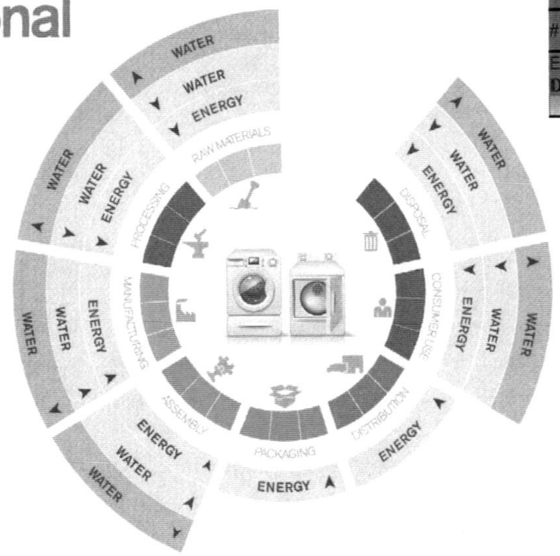

Functional Unit
One full load of laundry washed and dried

Geographic Scope
U.S., and Canada (distribution, use, and disposal) Europe (production)

Inputs
Energy
Water

Outputs
Water

home & personal care

working group

Baseline Models: **1**
Hotspots Identified: **6**
SPDs Identified: **16**
Example: **Laundry Detergent**

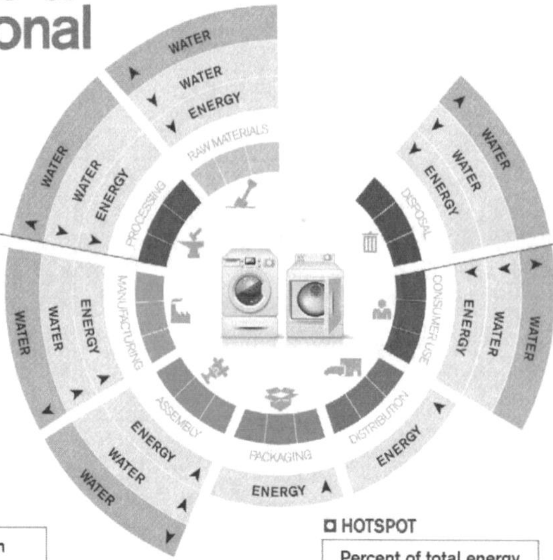

HOTSPOT

Percent of total energy in Ingredient Processing

SPD

1. Source of Feedstock (plant derived, oil/nat. gas derived)
2. Type of Ingredient (chemical, biological)
3. Production Efficiency

HOTSPOT

Percent of total energy in Mechanical Drying

SPD

1. Machine Efficiency
2. Residual Water in Clothes
3. Drying Time

HOTSPOT

Percent of total energy in Clothes Washing

SPD

1. Water Temperature
2. Machine Efficiency
3. Water per Load

Sector Working Groups

ELECTRONICS

Members: **19**
Participants: **25**
Launch Date: **Jan. 2010**
SMRS Prototype(s): **Laptop, Television**

Working Group Contact:
Carole Mars
Carole.mars@asu.edu

FOOD, BEVERAGE & AGRICULTURE

Members: **34**
Participants: **80**
Launch Date: **Jul. 2009**
SMRS Prototype(s): **Wheat Cereal, Flavored Yogurt, Fruit Juice**

Working Group Contact:
Sarah Lewis

HOME & PERSONAL CARE

Members: **27**
Participants: **78**
Launch Date: **Sept. 2009**
SMRS Prototype(s): **Laundry Detergent, Shampoo, Household Cleaner**

Working Group Contact:
Joby Carlson
Joby.carlson@asu.edu

Rising to Key Challenges

Consortium Working Groups (CWGs): Solving Key Challenges that Touch All Sectors

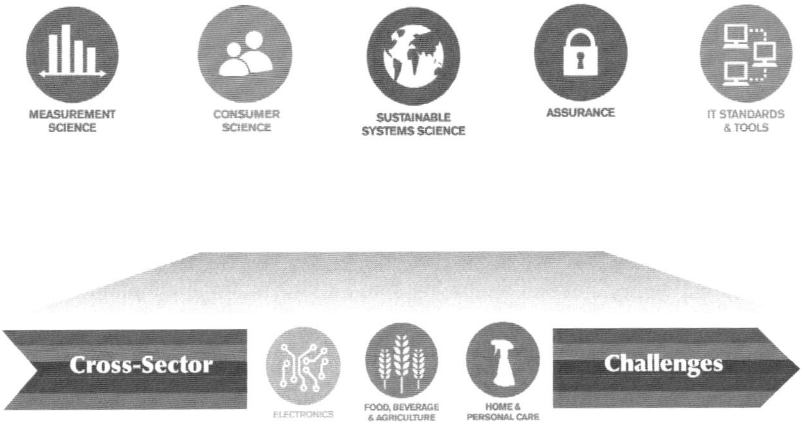

new earth

SUSTAINABILITY CONSORTIUM

Looking Ahead, Within TSC and Beyond

- Measurement Science Working Group: Divide and Conquer
- Social is Coming!
- Data: Filling the Black Holes
- Three More Messages

Meas. Sciences Working Group

- Draft SMRS Guidance Document
- Committees:
  - Life Cycle Impact Assessment
  - Baseline Model
  - Social Metrics

Challenges for Baseline Models

- "Warring database providers"
- Scaling: The need to create hundreds of baseline models, fast
- The need to get these models into *maximally efficient* open processes of review, improvement, finalization, & distribution

Summary of Approach in Consideration

- Systematized process
- Focus on the unit processes that matter
- Marketplace for best unit process data
- Store best unit processes in Library
- Deliver models for scrutiny, use
- Continually update and improve

The Core Tree Concept

The Core Tree

- Yes there are 5432 unit processes in the tree, but only 20-30 bring more than 1% of burden
- We don't have to ignore/ delete the little stuff; just replace micro-branches with **leaf nodes,** exposing what we could call "core branch nodes"

Threshold	Nodes
50 % (0.5)	5
10 % (0.1)	12
5 % (0.05)	15
1 % (0.01)	**31**
0.5 % (0.005)	43
0.1 % (0.001)	115
0.05 % (0.0005)	166
0.01 % (0.0001)	359
0.005 % (0.00005)	472
0.001 % (0.00001)	810
0 % (0)	5809

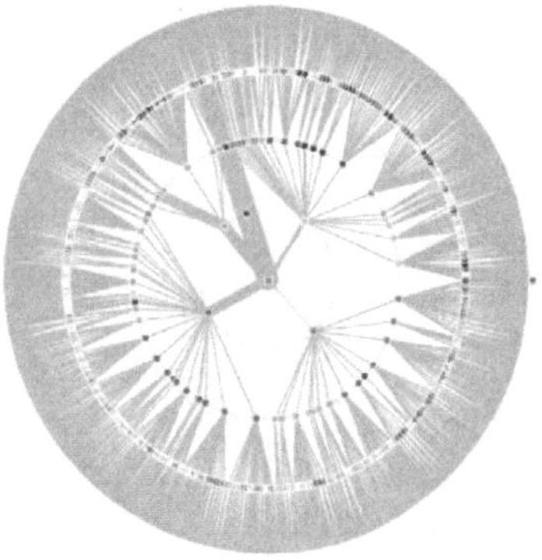

5809 nodes

31 nodes

Social is Coming!

- Not your Father's LCA
- UNEP/SETAC Project Group launched: "Life Cycle *Sustainability* Assessment"
- Industry Average Data Resource: Social Hot Spot Database
- Site- and Company-specific Social LCA of hot spots: In our (near) future

new earth

A Database for Social LCA

- Being applied to 7 product-level case studies within TSC
  - Done: Laptops, Orange Juice, and Laundry Detergent
  - April: Wheat Cereal, Shampoo, Strawberry Yogurt, tbd
- Web-based Open Transparent Data Portal
- www.socialhotspot.org
  ... check out the User Porta'

social hotspots database

new earth

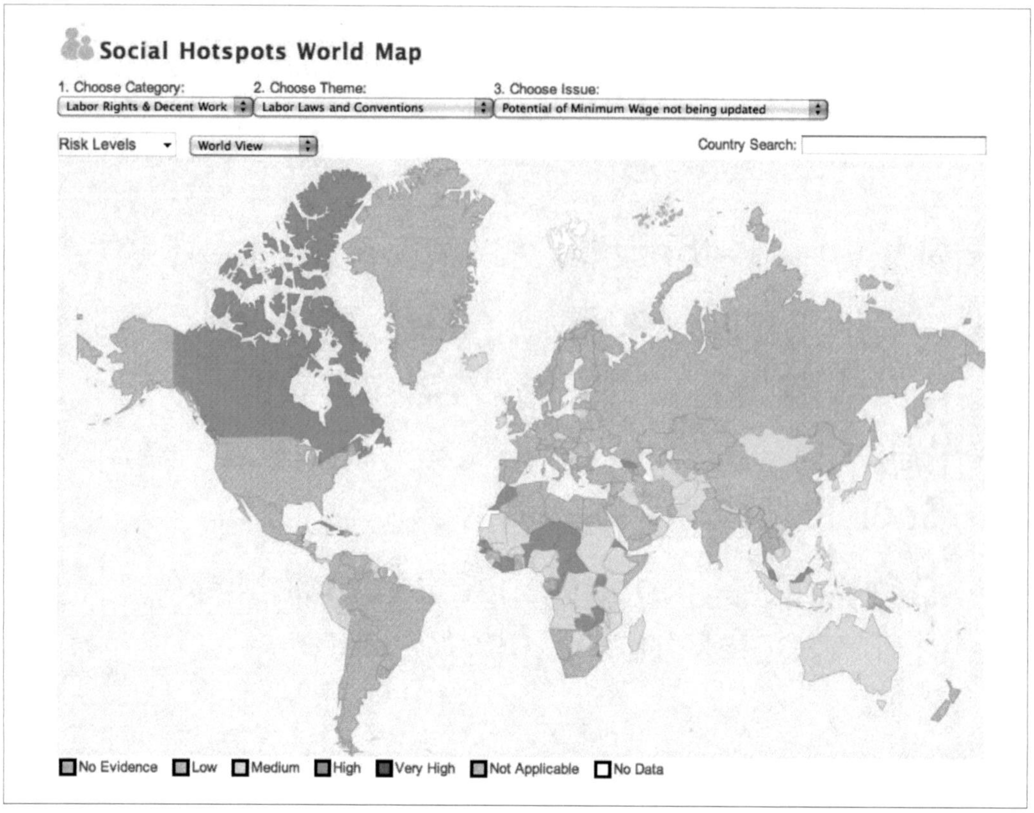
Social Hotspots World Map
1. Choose Category:
Labor Rights & Decent Work
2. Choose Theme:
Labor Laws and Conventions
3. Choose Issue:
Potential of Minimum Wage not being updated
Risk Levels
World View
Country Search:
No Evidence
Low
Medium
High
Very High
Not Applicable
No Data

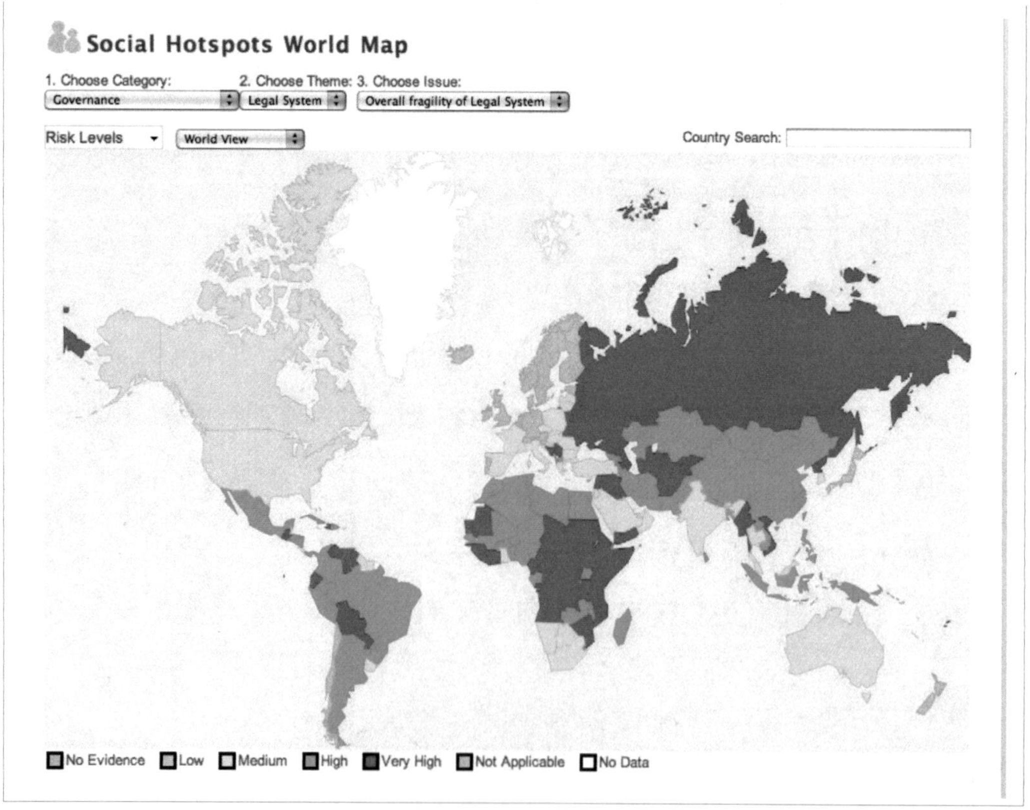
Social Hotspots World Map
1. Choose Category:
Governance
2. Choose Theme:
Legal System
3. Choose Issue:
Overall fragility of Legal System
Risk Levels
World View
Country Search:
No Evidence
Low
Medium
High
Very High
Not Applicable
No Data

China Social Risks and Opportunities
Potential of Minimum Wage not being updated
Potential of Country not passing Labor Laws
Potential of Country not adopting Labor Conventions
Potential of Average wage being < Minimum Wage
Potential of Average wage being < Non-poverty Guideline
Percent of Population living on <$2/day
Risk of Child Labor
Risk of Forced Labor
Risk of Population working >48 hours/week
Risk of not having Freedom of Association Rights
Risk of not having Collective Bargaining Rights
Risk of not having the Right to Strike
Overall fragility of Legal System
Number of indigenous population
Risk of country not adopting Indigenous ILO convention and UN Declaration
Risk of country not passing Laws to protect indigenous
Potential for High Conflict
Overall fragility of Gender Equity
Number of Children out of School – male
Number of Children out of School – female
Number of Children out of School – total
Risk of not having access to Improved Drinking Water – rural
Risk of not having access to Improved Drinking Water –urban
Risk of not having access to Improved Drinking Water – total
Risk of not having access to Improved Sanitation – rural
Risk of not having access to Improved Sanitation – urban
Risk of not having access to Improved Sanitation – total
Risk of not having access to a Hospital Bed
Low
Medium
High
Very High

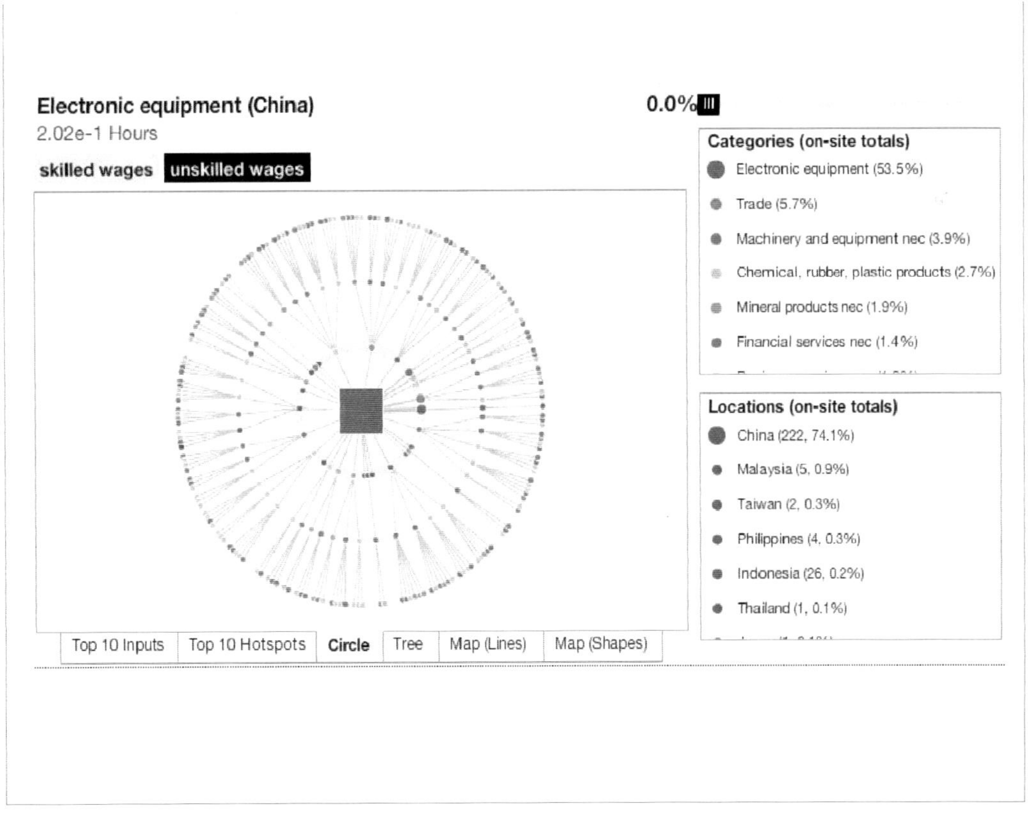
Electronic equipment (China)
2.02e-1 Hours
skilled wages
unskilled wages
0.0%
Categories (on-site totals)
Electronic equipment (53.5%)
Trade (5.7%)
Machinery and equipment nec (3.9%)
Chemical, rubber, plastic products (2.7%)
Mineral products nec (1.9%)
Financial services nec (1.4%)
Locations (on-site totals)
China (222, 74.1%)
Malaysia (5, 0.9%)
Taiwan (2, 0.3%)
Philippines (4, 0.3%)
Indonesia (26, 0.2%)
Thailand (1, 0.1%)
Top 10 Inputs
Top 10 Hotspots
Circle
Tree
Map (Lines)
Map (Shapes)

Data: Filling the Black Holes

- UNEP/SETAC Global Guidance for Life Cycle Inventory Data: Japan, Feb 2011
- Future Scenarios Group identified two technology innovations which can bring major increase in LCI data availability:
  - Making Un-tapped National and Central Databases Easy to Use in Creating LCI data
  - Empowering/Incentivizing Bottom-up Data Creation

new earth

Making un-tapped Data Resources Harvestable

- USDA pilot project coming to completion
  - Using existing federal resources to develop region-specific unit processes for US agriculture
- Field-To-Market data development project is another good example of this approach
- EPA hosting internal workshop April
- Making this scalable, and global:
  - Governments in US and UK making data resources available as Linked Open Data: speeds conversion to LCI

new earth

Empowering / Incentivizing Bottom-up Data Creation

- No cost barrier to entry for Sustainability Footprinting
- Economic Value > Cost of Use
- Aggregate unit process data from 3 or more users, under confidentiality
  - Provide benchmarks to users
  - Provide generic data to world

earthster

new earth

Three More Messages

- Innovation
- Innovation
- Innovation

new earth

Innovation 1: Supply Side Macro

- Sector Working Groups,with other talent, putting a baseline model to work to identify system-level innovation opportunities

new earth

Innovation 2: Supply Side Micro

- Individual companies, using tools such as Earthster:
  - Use local assessment to query "Match Maker" database
  - Identify most promising and appropriate resources for profitable sustainable innovation
  - Document progress in ways that speed spread of proven innovations

new earth

Innovation 3: Demand Side

WORLD ECONOMIC FORUM

COMMITTED TO IMPROVING THE STATE OF THE WORLD

"Rethinking Growth"

"From Incrementalism
To Transformation"

The Consumption Dilemma

Leverage Points for Accelerating Sustainable Growth

Innovation 3: Demand Side

- Imagine promoting sharing of innovation among citizens:
  - Maximizing well-being
  - Ensuring our "handprints" are larger than our footprints

new earth

Initiatives impacting **packaging**

presented by Laura Rowell, Director,
Sustainable Packaging, MWV
March 17, 2011

MWV Sustainability Leadership

Industry Initiatives/Leadership

- Global Packaging Project (Consumer Goods Forum)
- ISO Packaging and the Environment Standards
- Keystone Center's Green Products Roundtable
- AMERIPEN founding member
- Sustainable Packaging Coalition founding member

SUSTAINABLE PACKAGING COALITION

Dow Jones Sustainability Indexes

Recognition

- Carbon Disclosure Project / Supply Chain Collaboration
  - Cited for leadership and performance 2010
- Dow Jones Sustainability Index
  - Listed for past 7 years; containers and packaging leader for 6 years
- Ranked #161 on Newsweek's 500 Greenest US Companies

MWV

Global Voluntary Initiatives

In addition to legislative challenges, sustainable packaging initiatives are also being driven by organizations not on the legislative/regulatory radar

This is being supported by retailers and brand owners who feel they can act more quickly, and with more positive results, than governments

Compliance is voluntary, but expected, and will impact:

- How we DO what we do
- How we MEASURE what we do
- How we TALK about what we do

3 MWV

Players

Global CEO Forum

- Organized Global Packaging Project at request of Tesco and Unilever CEOs
- Charged with development of sustainable packaging metrics and guidelines

Consumer Goods Forum

- Merger of CIES, Global CEO Forum and the Global Commerce Initiative (GCI)
- Co-chaired by CEOs of Coca Cola and Carrefour
- Has assumed responsibility for the Global Packaging Project

Sustainable Packaging Coalition

- Established in 2002
- A project of GreenBlue – a non-profit organization founded in 2002 through McDonough Braungart Design Chemistry (MBDC)
- Along with EUROPEN, provides staff support and consulting to Global Packaging Project

4 MWV

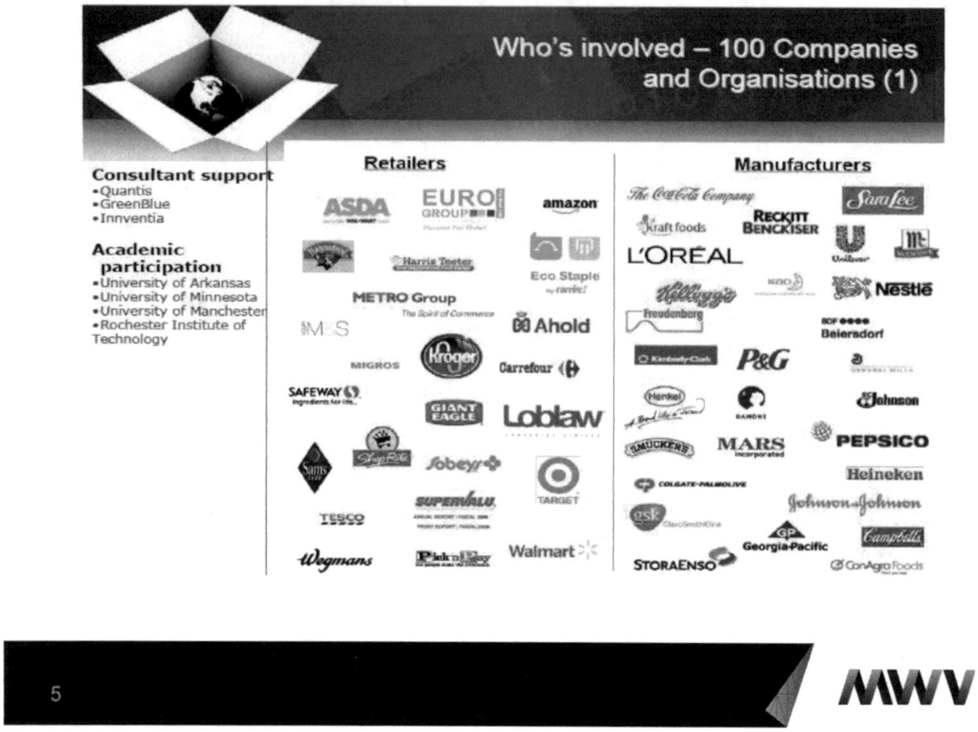
Who's involved – 100 Companies and Organisations (1)
Consultant support
• Quantis
• GreenBlue
• Innventia
Academic participation
• University of Arkansas
• University of Minnesota
• University of Manchester
• Rochester Institute of Technology
Retailers
Manufacturers
5
MWV

Who's involved – 100 Companies and Organisations (2)
Associations
Packaging manufacturers
6
MWV

Global Packaging Project

Underlying concept: Development of common definitions and metrics, used and communicated in compliance with applicable international standards, will avoid unintended burden shifting

Approach

- Draft metrics and methodologies originally drawn from SPC
  - Social
  - Environment (attribute and LCA)
  - Economic
- Alignment with globally recognized standards and measurements
- Test metrics through pilot projects

Impact on packaging industry

- Avoid multiple approaches to scorecards and metrics; uniform scope and methodologies
- Allows everyone to speak the same language

7 MWV

Global Packaging Project

Overall an excellent (but sometimes a bit confusing) document

Results

- Good clarity that choice of metrics (and amount of them) depends on business interests and goals
  - However, this may not be an option for packaging suppliers
- 42 metrics, covering:
  - Social – 13 (most are a checklist of policies relating to workplace practices)
  - Environment – 11 plus 2 on performance of EMS and energy audits
  - LCA indicators and attributes – 14
  - Economic – 2

8 MWV

ISO Standards

Environmental Packaging Standards

- Goal to transpose the EU CEN Standards to global ISO Standards

These cover

- EN 13427 – Umbrella Standard
- EN 13428 – Optimization by Source Reduction (and affiliated reports)
- EN 13429 – Reuse
- EN 13430 – Material Recovery
- EN 13431 – Energy Recovery
- EN 13432 – Organic Recovery
- Addition of Technical Report on Chemical Recovery

Impact on paperboard packaging industry

- Global approach to standards; discourage development of proscriptive requirements
- Supports Global Packaging Project metrics and methodologies

MWV

redipoint
&
Packaging Scorecard

Presented by
Victor Bell – EPI
Charles Walsh - ECRM

SUSTAINABLE PACKAGING SYMPOSIUM 2011

The redipoint Partnership

EPI
- Specialists in global environmental pkg & product stewardship requirements
- Provides regulatory tracking and other compliance services
- Customers include;

ECRM
- Provides Business Solutions by integrating Process, Vision and Technology
- Provides solutions for reporting & scorecarding requirements

2

The redipoint Solution: Overview

redipoint offers an innovative solution for worldwide compliance reporting, corporate benchmarking and data management needs, including:

- Extended Producer Responsibility (packaging, electronics and battery fee reports to authorities and waste recovery organizations and submission of reports to support worldwide compliance with environmental legislation)
- Allowing suppliers to report directly into retailer (Walmart) sustainable packaging scorecards
- GPP Tool
- Corporate environmental benchmarking
- Comprehensive Bill of Materials (BOM) data management
- Design-for-Environment analyses for evaluating packaging and product designs and prototypes materials impacts

Introduction to redipoint

- redipoint greatly simplifies BOM management, providing companies a single site to manage product and packaging data across divisions and around the world.
- Users can enter and edit BOM data for all items, either manually or by bulk upload.
- Once entered, BOM data can be used to automatically populate fee reports and estimates, export for retailer scorecards, etc.
- Developed in partnership by EPI and ECRM.

4

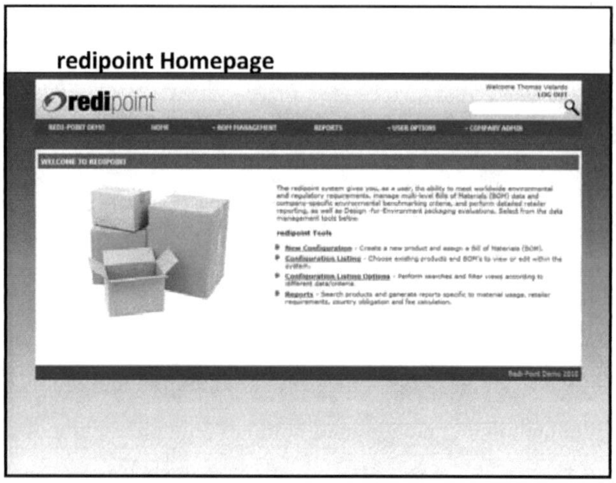
redipoint Homepage
redipoint

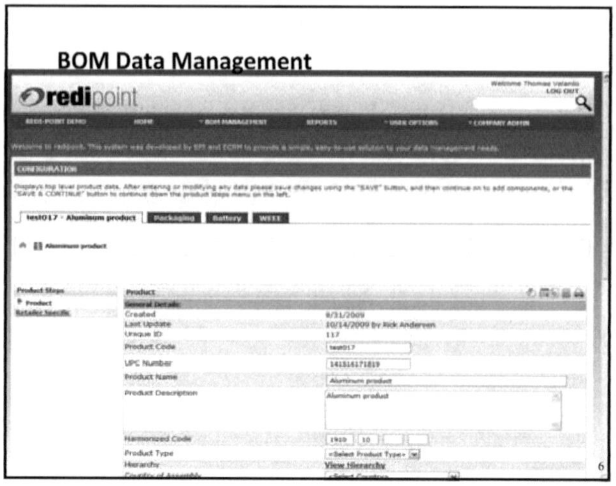
BOM Data Management
redipoint
6

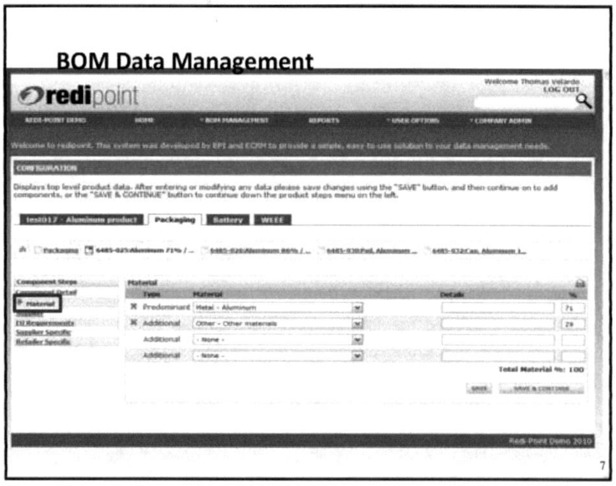
BOM Data Management
redipoint
7

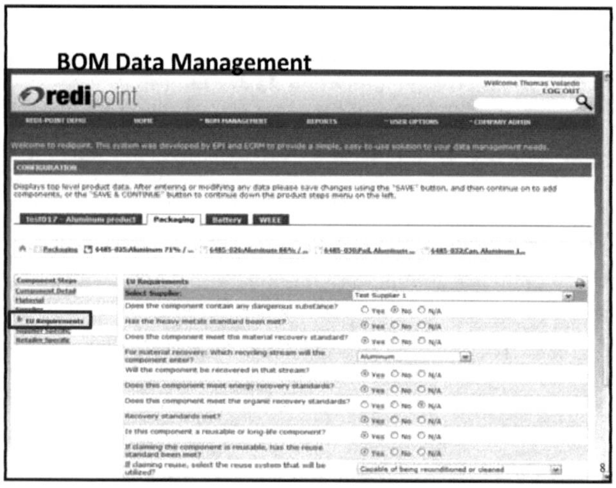
BOM Data Management
redipoint
8

Reports

- Redipoint features a powerful reporting component that enables users to generate customizable reports on demand.
- Six different report types are available:
  - Material Usage
  - Retailer Report
  - Obligation
  - Fee Calculator
  - Metric Report
  - Retailer Scorecard Calculator

9

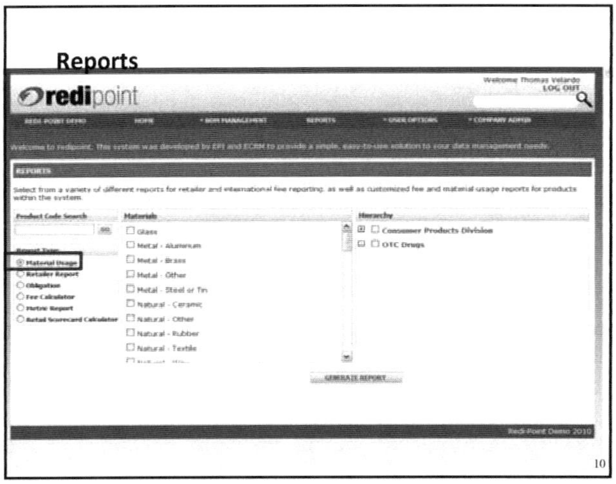

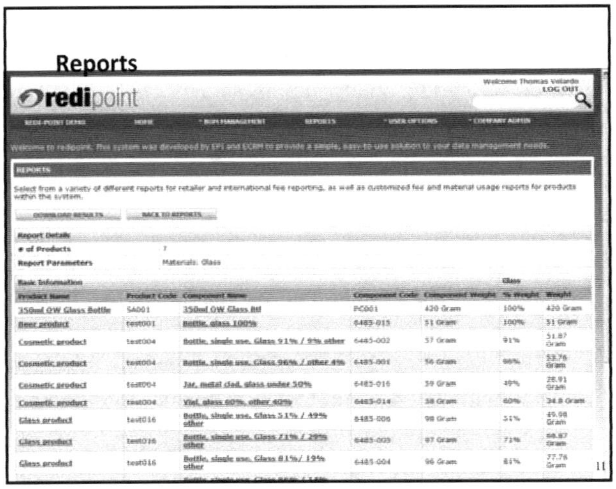

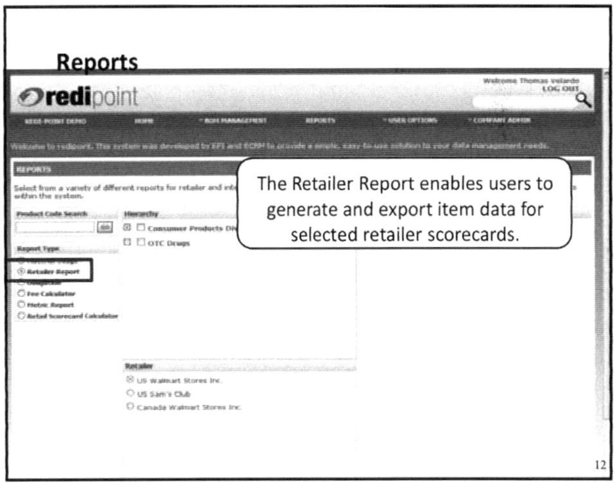

www.enviro-pac.com

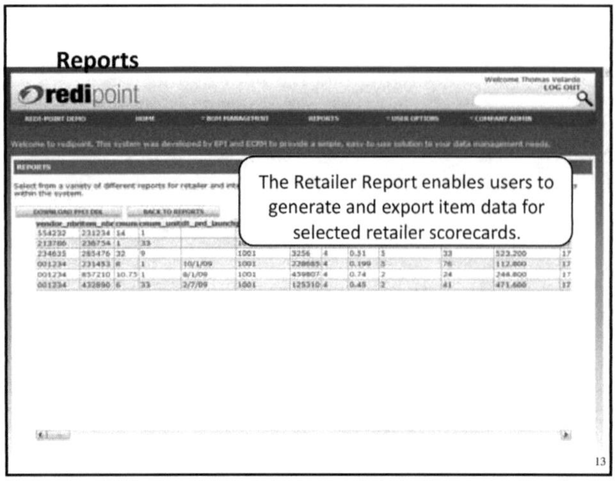
Reports
redipoint
The Retailer Report enables users to generate and export item data for selected retailer scorecards.
13

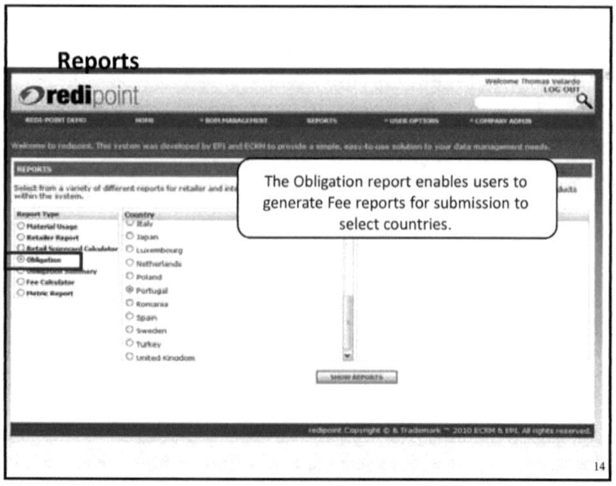
Reports
redipoint
The Obligation report enables users to generate Fee reports for submission to select countries.
14

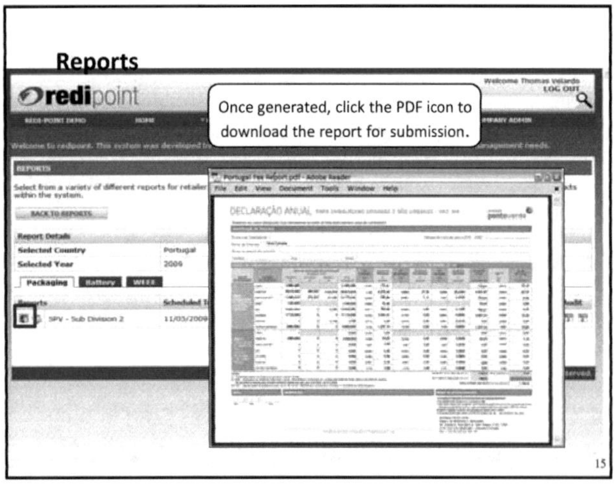
Reports
redipoint
Once generated, click the PDF icon to download the report for submission.
15

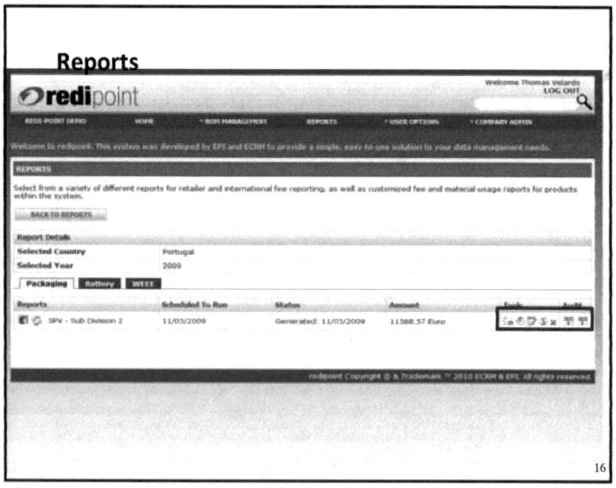
Reports
redipoint
16

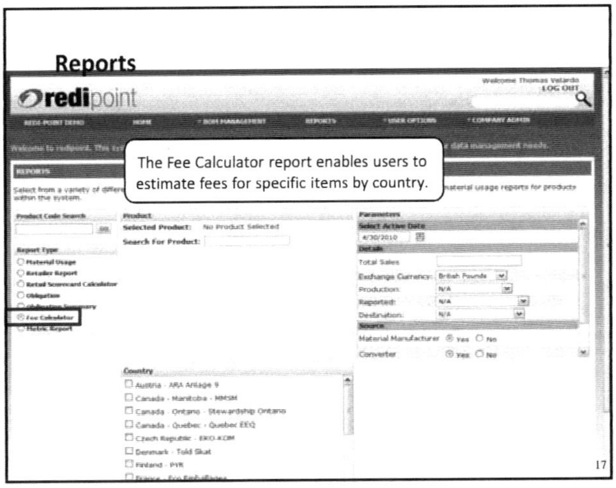
Reports
redipoint
The Fee Calculator report enables users to estimate fees for specific items by country.
17

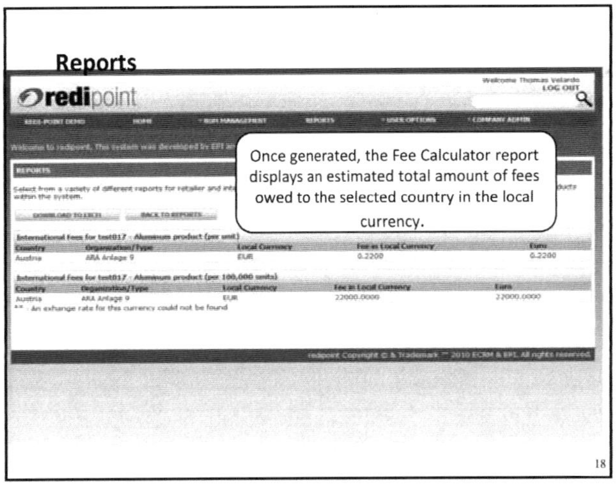
Reports
redipoint
Once generated, the Fee Calculator report displays an estimated total amount of fees owed to the selected country in the local currency.
18

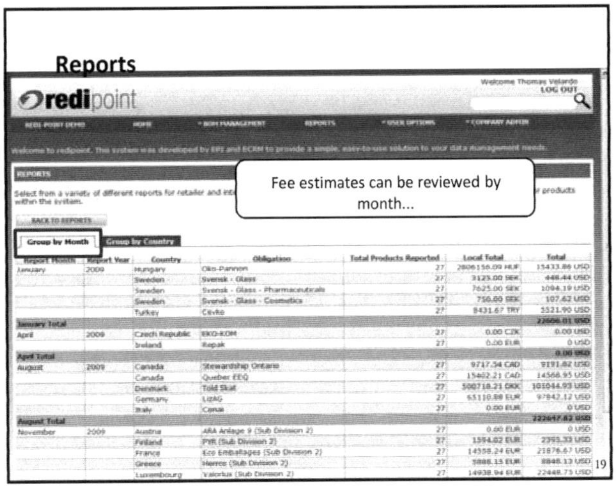
Reports
redipoint
Fee estimates can be reviewed by month…
Group by Month
Group by Country
19

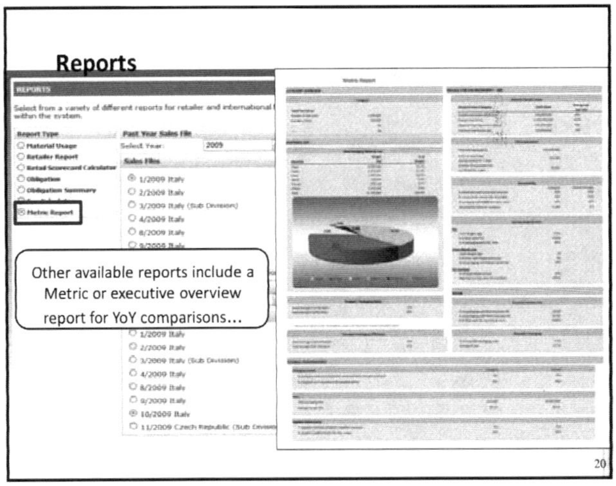
Reports
Other available reports include a Metric or executive overview report for YoY comparisons…
20

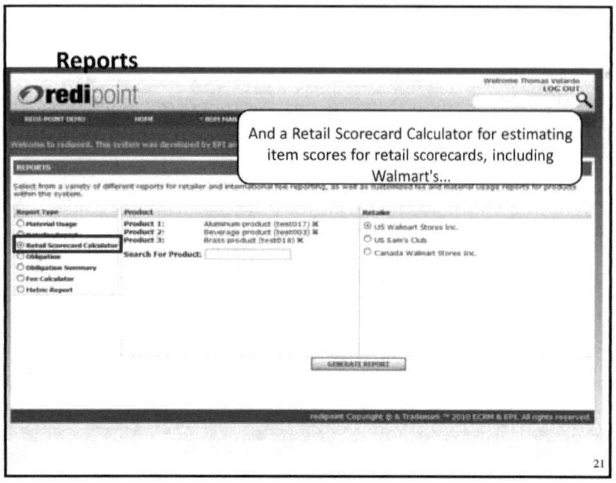
Reports
redipoint
And a Retail Scorecard Calculator for estimating item scores for retail scorecards, including Walmart's...
21

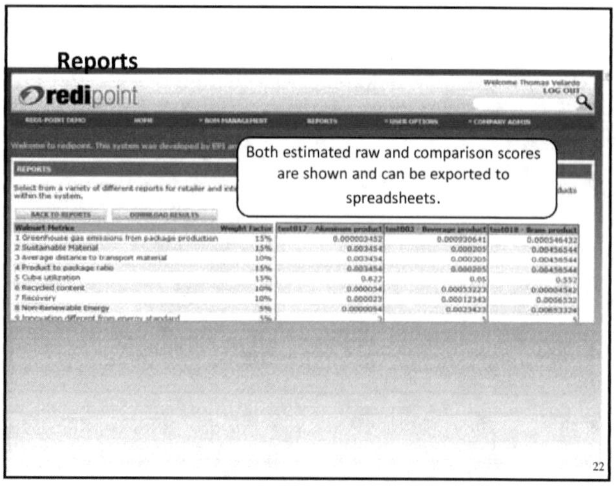
Reports
redipoint
Both estimated raw and comparison scores are shown and can be exported to spreadsheets.
22

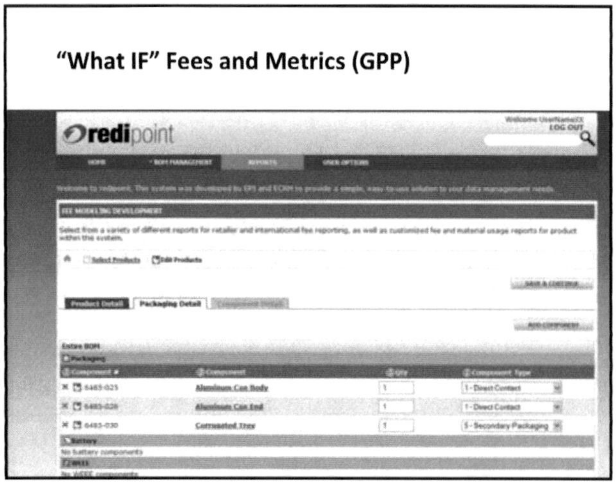
"What IF" Fees and Metrics (GPP)
redipoint

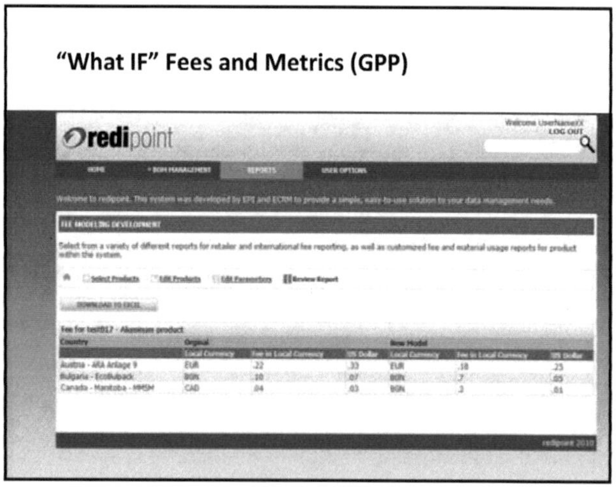
"What IF" Fees and Metrics (GPP)
redipoint

ECRM Packaging Scorecard

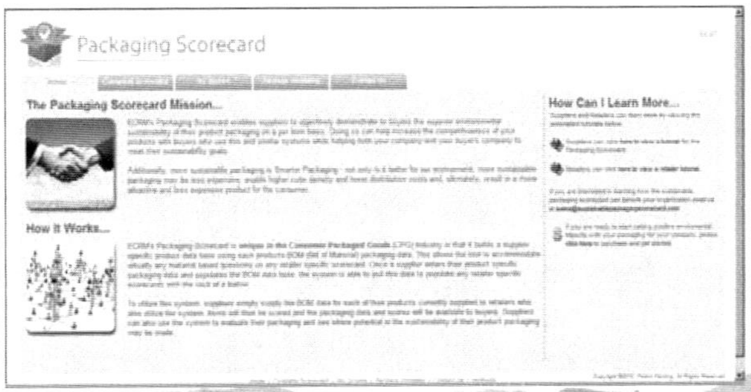

Packaging Scorecard Background

- In 2006 ECRM, in conjunction with Walmart and its Packaging Sustainable Value Network, developed the software to support the Walmart Sustainable Packaging Scorecard
- All questions, material metrics and metric calculations used in Walmart's Scorecard were developed by the retailer's Packaging SVN
- The Packaging SVN is comprised of more than 200 representatives from the Packaging Industry, CPG companies, Trade Associations, Academics, Government Organizations and Non GO's
- Walmart's Scorecard is currently in use in the United States, Canada and Mexico with ongoing international rollouts planned and ECRM continues to support Walmart in this rollout

What is the ECRM Packaging Scorecard

- ECRM Packaging Scorecard uses the same metrics and metric calculations as in the Walmart Packaging Scorecard
- As **GPP** identifies new attributes and metrics, many of these will be incorporated into the Packaging Scorecard
- ECRM's Packaging Scorecard makes it easy for retailers to compare packaging sustainability by item and by supplier
- The Packaging Scorecard is a solution that is available to all retailers globally and is capable of measuring a complete set of metrics, including Green House Gas emissions, for any packaged goods product

Benefits of the ECRM Packaging Scorecard

- The Packaging Scorecard may be used by Retailers seeking improvements in the sustainability of their product packaging
- The Scorecard can be accessed by Retailers supplier's directly through ECRM Hosted Website
- Suppliers enter their Product Packaging Data one time and it may be captured for all Retailers who utilize the ECRM Packaging Scorecard

Benefits of the ECRM Packaging Scorecard

- Allows Retailers to easily compare packaging sustainability across a product category
- Retailers can use this data to make more informed purchasing decisions and source products with more sustainable packaging
- Eliminates the redundancy of suppliers entering the same product packaging data multiple times for multiple customers

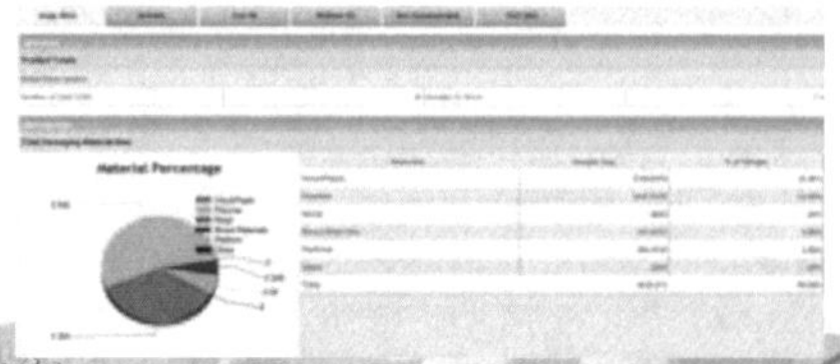

29

Packaging Scorecard - How it works

- ECRM's Packaging Scorecard is a web based subscription solution in a secured environment to ensure the highest level of data security / integrity
- Suppliers access the Packaging Scorecard Website to enter in their Product Packaging by SKU
- Suppliers have a variety of entry options including single item and bulk upload

This information is copyrighted and cannot be copied or distributed without prior written consent from Environmental Packaging International

30

Packaging Scorecard - How it works

- Suppliers are able to use the Package Modeling feature to model ways to improve their products packaging sustainability
- Suppliers are able to dynamically access Packaging Suppliers who are members of the Greener Package Database directly through the Package Modeling tool

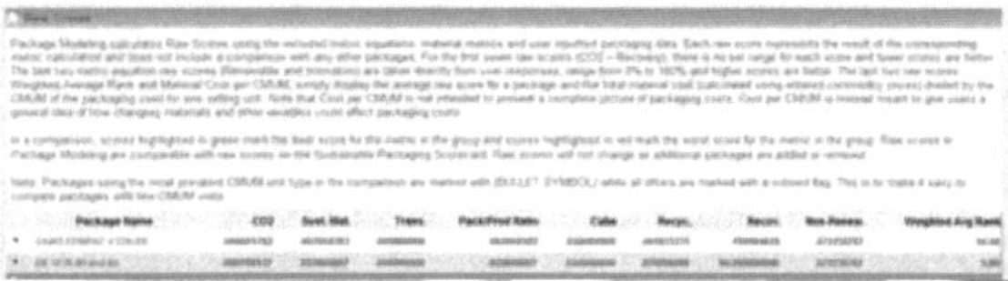

This information is copyrighted and cannot be copied or distributed without prior written consent from Environmental Packaging International

31

Packaging Scorecard - How it works

- Packaging Scores and Reports are generated automatically using standardized reporting tools including;
  - YOY changes
  - Supplier Roll Up
  - Category, Department or Buyer level comparisons
- Aggregated Sales data feeds may also be accepted in order to generate aggregate level measurements

This information is copyrighted and cannot be copied or distributed without prior written consent from Environmental Packaging International

32

Packaging Scorecard - Scoring

- As in the Walmart Scorecard, our scorecard currently measures the following;
  - Greenhouse Gas Emissions
  - Sustainable Material
  - Average Distance to Transport
  - Package to Product Ratios
  - Cube Utilization
  - Recycled Content
  - Recovery
  - Renewable Energy
  - Innovation
- As the **GPP metrics** are finalized the scorecard will be aligned with them

33

Guidelines to Sustainability Claims

- Introducing Greener Package Guidelines to Sustainability Claims
- Picks up where FTC green guidelines leave off
- 20-page PDF written by EPI
- This is the standard against which reviewers review claims
- Download for free at GreenerPackage.com/guidelines
- Living document! Submit feedback!

Who put this together?

- 40+ members of Greener Package advisory board (end users + suppliers)
- Months spent on packaging sustainability questions, guided by independent consultancy EPI.
- Basically: your peers plus experts!
- Worked hard to achieve broad consensus

Database: standard questions & answers

Post-consumer recycled content. This material or package contains a minimum post-consumer recycled content in the following amount(s):
- ☐ 10 - 24%
- ☑ 25 - 49%
- ☐ 50 - 74%
- ☐ 75 - 99%
- ☐ 100%

- Questions asked and answered the same way for everyone
- 12 questions pertaining to sustainability claims.
- And if you answer YES to voluntary third-party review…

THIRD-PARTY REVIEW

Do you wish to have a third party review the claims you make for the product? (More info)
- ◉ Yes
- ○ No

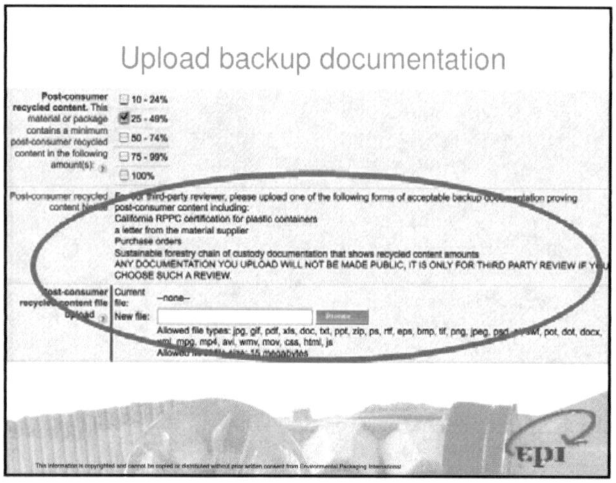

What the reviewer does

- Are your papers in order?
- Do you have documentation backing up your claim?
- Does it match the requirements set out in the Guidelines to Sustainability Claims?
- In short, are these claims supportable and made in good faith?

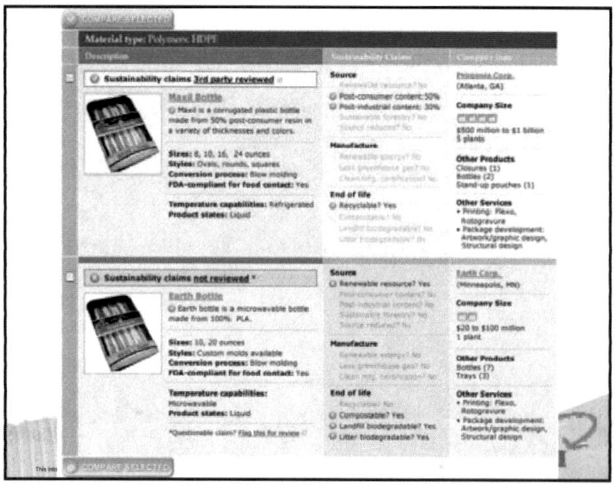

Contact Information

Victor Bell- EPI
401-423-2225
vbell@enviro-pac.com

Charles Walsh - ECRM
479-464-0564
cwalsh@ecrm.marketgate.com

www.redi-point.com
www.sustainablepackagingscorecard.com

COMPASS®
COMPARATIVE PACKAGING ASSESSMENT

Measurements, Data & Analysis

Minal T. Mistry

Sustainable Packaging Symposium
18 March 2011

SUSTAINABLE PACKAGING COALITION®

a project of GreenBlue

GreenBlue®

A non-profit organization focused on engaging industry to advance sustainability initiatives.

- Sustainable Packaging Coalition® (SPC)
- Paper and Forestry (MetaFore)
- Closing the Loop
- Advisory Services
- Tools: CleanGredients®, EPAT®, COMPASS®

www.greenblue.org | www.sustainanblepackaging.org | www.cleangredients.org | www.metafore.org

SUSTAINABLE PACKAGING COALITION®

a project of GreenBlue

COMPASS®
COMPARATIVE PACKAGING ASSESSMENT

a **design-phase** web application that provides comparative **environmental profiles** of **packaging** alternatives based on **life cycle** assessment **metrics** and **attributes**

SUSTAINABLE PACKAGING COALITION®

a project of GreenBlue

Redefining Industrial Design for Packaging

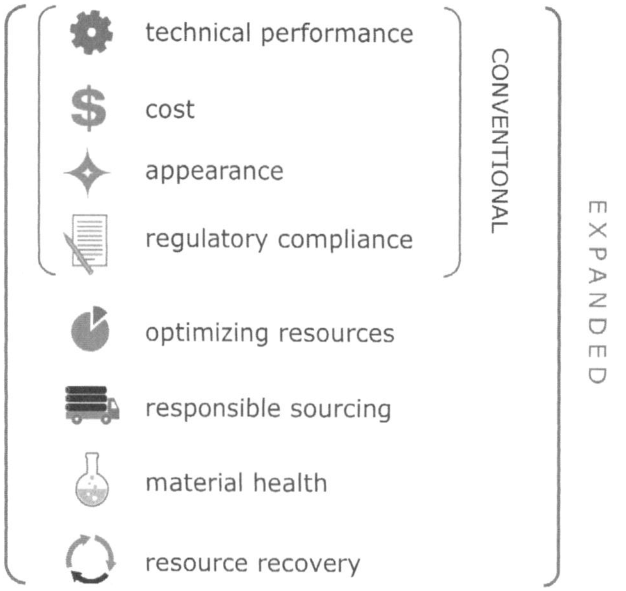

SUSTAINABLE PACKAGING COALITION®

a project of GreenBlue

Influence of Design (downstream →)

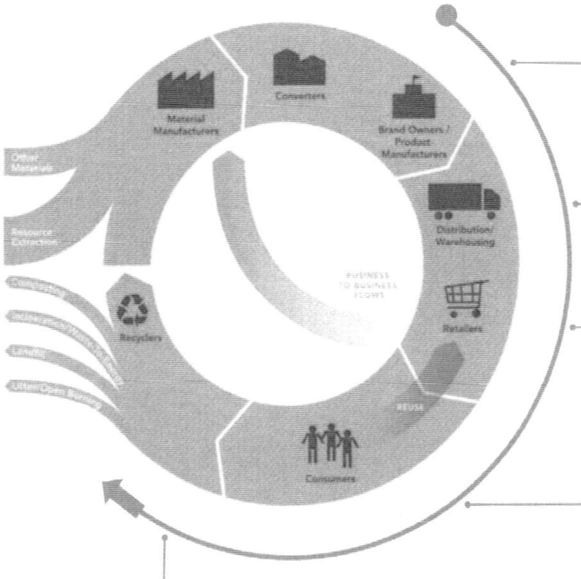

BRANDS
- image
- perception

DISTRIBUTION
- transport efficiency
- product loss

RETAIL
- on shelf differentiation
- shelf life
- safety

CONSUMER
- product appeal
- safety
- satisfaction

WASTE DISCARD
- effective recovery
- reduce disposal

SUSTAINABLE PACKAGING COALITION®

a project of GreenBlue

Influence of Design (← upstream)

MARKET SIGNALS
- demand for sustainable sourcing
- reduced energy intensity
- reduced impacts

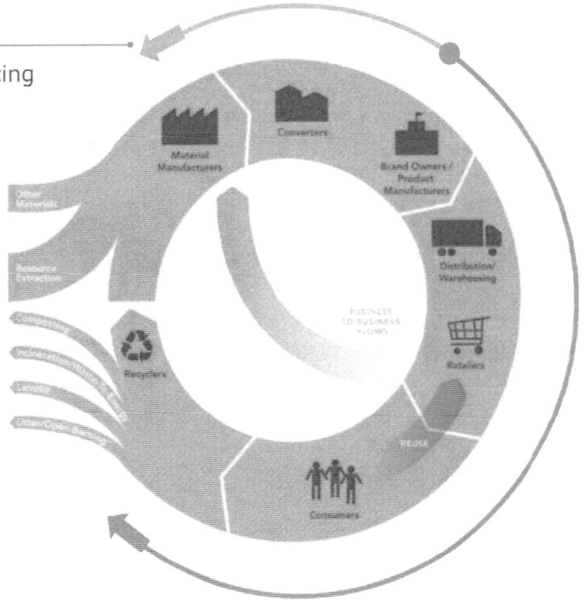

SUSTAINABLE PACKAGING COALITION®

a project of GreenBlue

Market Trend

Post Market Eco-Footprint or Score

SUSTAINABLE PACKAGING COALITION®

a project of GreenBlue

Design Phase Guidance

Benchmark Packaging Portfolio

SUSTAINABLE PACKAGING COALITION®

a project of GreenBlue

© 2011

Design Phase Guidance

Inform New Design

CONSENSUS BASED DEVELOPMENT

Actors

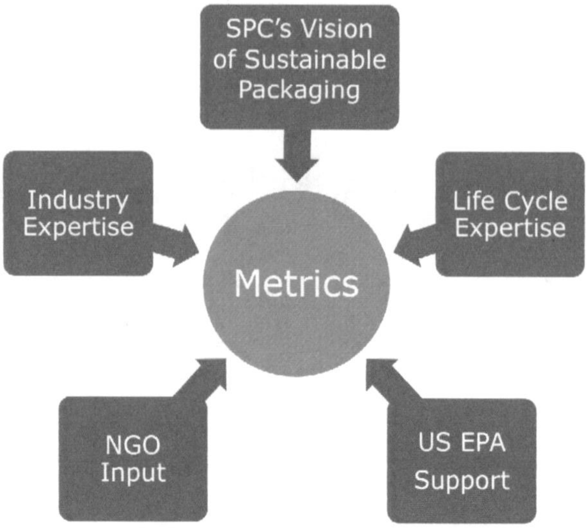

SUSTAINABLE PACKAGING COALITION®

a project of GreenBlue

Data Model Vetted and Verified

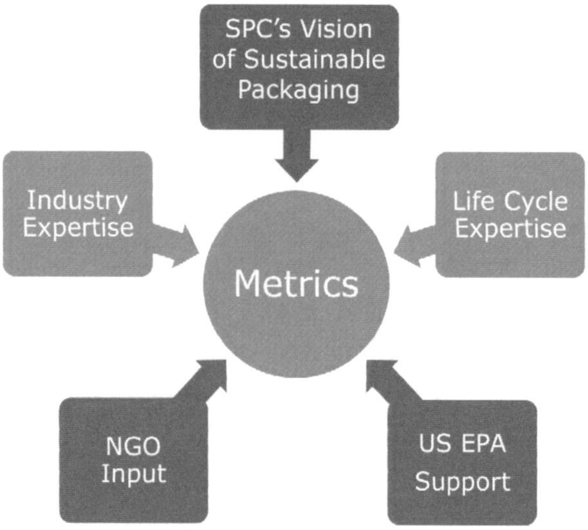

SUSTAINABLE PACKAGING COALITION®

a project of GreenBlue

Metrics Relevant to Packaging

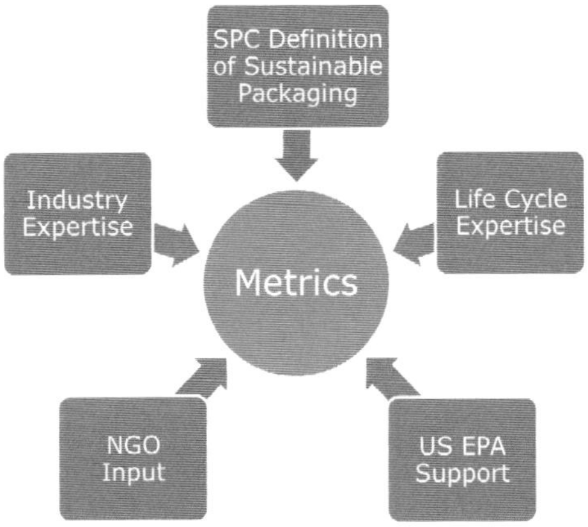

CONSUMPTION METRICS
- FOSSIL FUEL
- WATER
- BIOTIC RESOURCES
- MINERAL RESOURCES

EMISSION METRICS
- GREENHOUSE GASES
- HUMAN HEALTH
- AQUATIC TOXICITY
- EUTROPHICATION

PACKAGING ATTRIBUTES
- CONTENT
- SOURCE
- SOLID WASTE

MATERIAL HEALTH

SUSTAINABLE PACKAGING COALITION®

a project of GreenBlue

Life Cycle Model in COMPASS

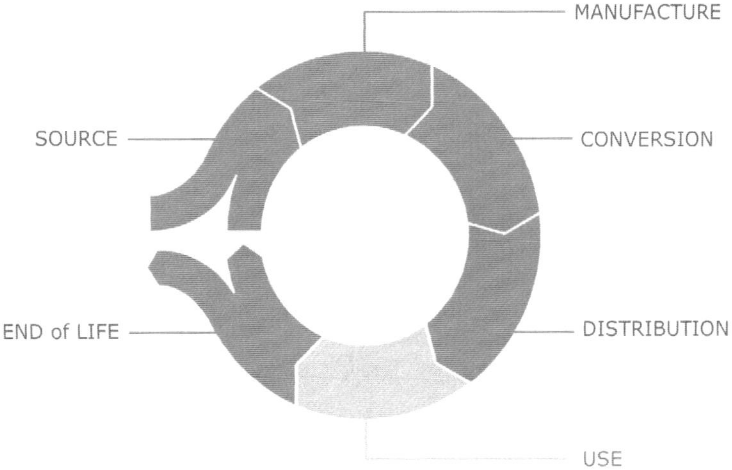

SUSTAINABLE PACKAGING COALITION®

a project of GreenBlue

Build Scenarios Using Components

COMPONENTS **PRIMARY PACKAGE** **SECONDARY PACKAGE**

SUSTAINABLE PACKAGING COALITION®

a project of GreenBlue

Packaging System

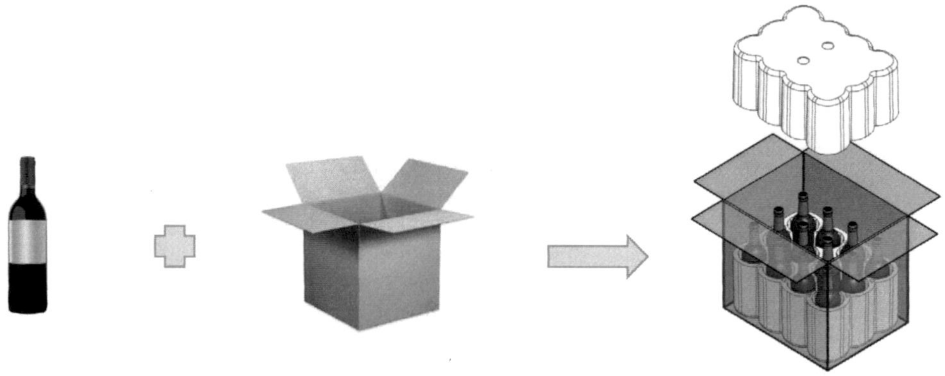

PRIMARY PACKAGE **SECONDARY PACKAGE** **PACKAGING SYSTEM**

Add distribution related transport for components, packages and shipping the system to the DC

SUSTAINABLE PACKAGING COALITION®

a project of GreenBlue

IMPACT ASSESSMENT

SUSTAINABLE PACKAGING COALITION®

a project of GreenBlue

Consistent Modeling

- Regions represented:
  - U.S.
  - Canada
  - Europe
- Materials
  - Uses consistent boundary for cradle-to-gate data
  - Data checked for gaps or omissions
- Conversion processes – compatible for packaging applications
- Distribution – regionally available vehicle options
- End-of-life – regional packaging disposal and recovery rates factored in
- Life cycle calculations checked for double counting concerns
- Current data sets vetted by industry and LCA experts

SUSTAINABLE PACKAGING COALITION®

a project of GreenBlue

Easy to Learn User Interface

Distribution and End-of-Life

Primary Packages

Secondary Package

Packaging System

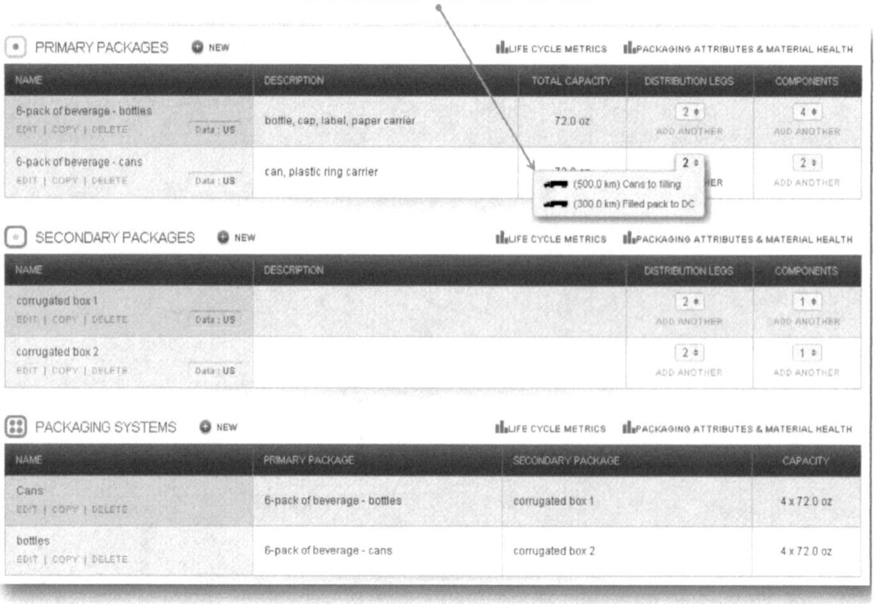

SUSTAINABLE PACKAGING COALITION®

a project of GreenBlue

Assess Component Contribution

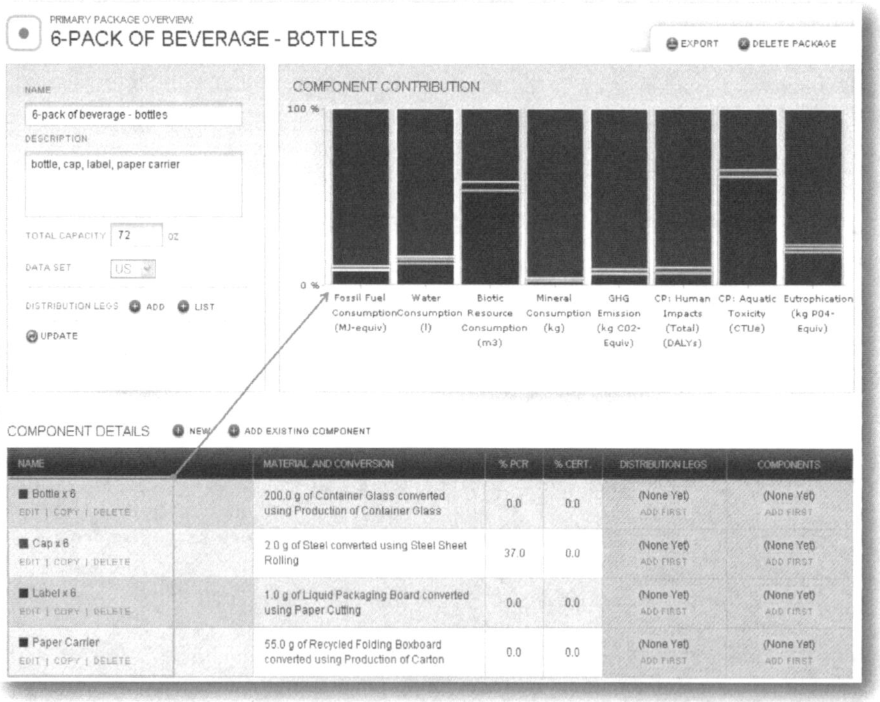

SUSTAINABLE PACKAGING COALITION®

a project of GreenBlue

Full Project Spreadsheet Export

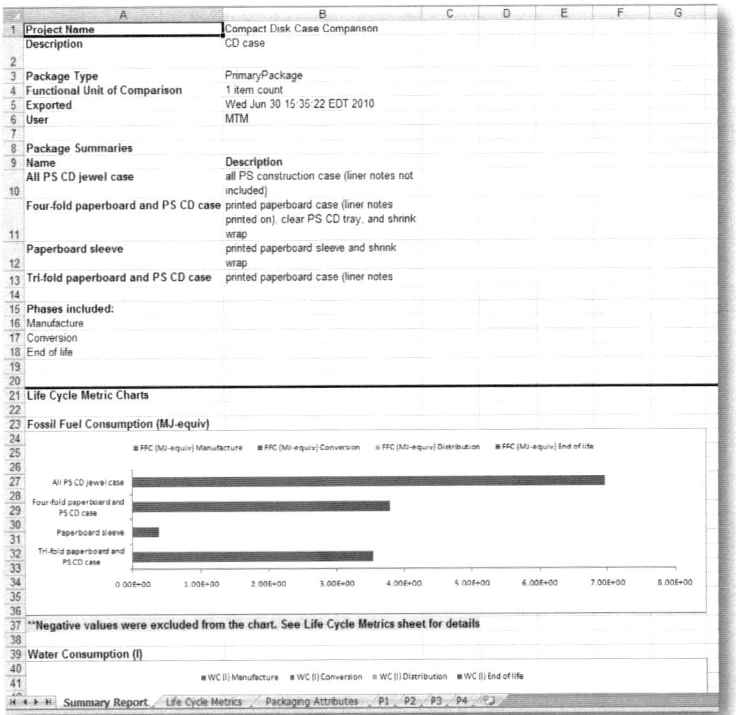

SUSTAINABLE PACKAGING COALITION®

a project of GreenBlue

Comparative Assessment

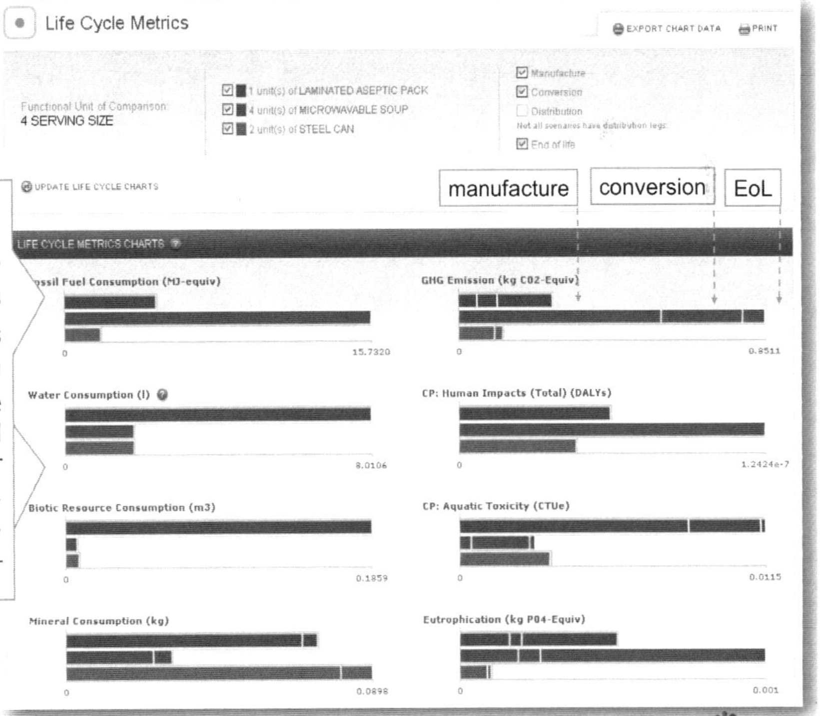

- Compare design alternatives
- Test scenarios
- Apples to Apples comparison
- Life cycle consumption and emissions metrics for manufacture, conversion, distribution, and EoL

SUSTAINABLE PACKAGING COALITION®

a project of GreenBlue

Packaging Attributes & Material Health

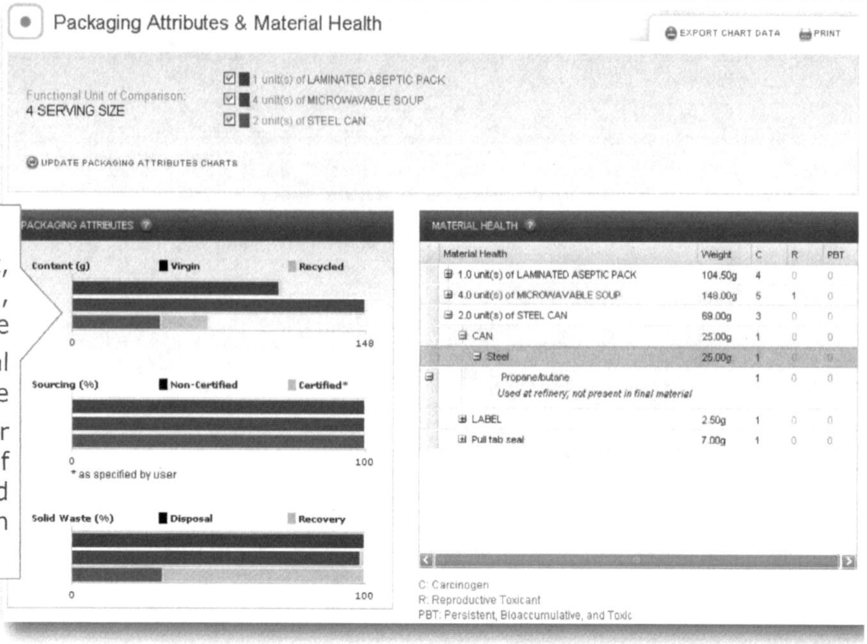

- Track PCR content, source certification, and solid waste
- Review material health profile
- Export data for project chain of custody and communication

SUSTAINABLE PACKAGING COALITION®

a project of GreenBlue

GLOBAL PACKAGING PROJECT

SUSTAINABLE PACKAGING COALITION®

a project of GreenBlue

GPP Life Cycle Metrics

- Cumulative energy demand
- Fresh water consumption
- Land use

- Global warming potential (GWP)
- Ozone depletion
- Toxicity, cancer
- Toxicity, non cancer
- Particulate respiratory effects
- Ionizing radiation (human)
- Photochemical ozone creation potential (POCP)
- Acidification potential
- Aquatic eutrophication
- Freshwater ecotoxicity potential
- Non-renewable resource depletion

SUSTAINABLE PACKAGING COALITION®

a project of GreenBlue

COMPASS Life Cycle Metrics

Consumption

- Fossil Fuel
- Water
- Biotic Resources
- Mineral

Emission

- Greenhouse Gases
- Human Health
  - Particulates
  - Carcinogens
  - Non-carcinogenic toxicants
- Aquatic Toxicity
- Eutrophication

SUSTAINABLE PACKAGING COALITION®

a project of GreenBlue

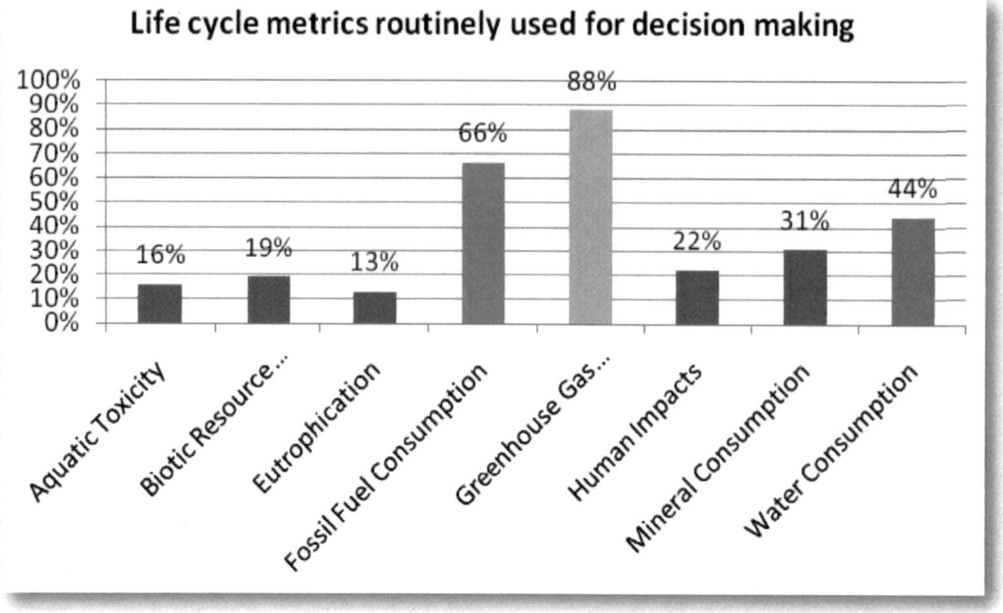

SUSTAINABLE PACKAGING COALITION®

a project of GreenBlue

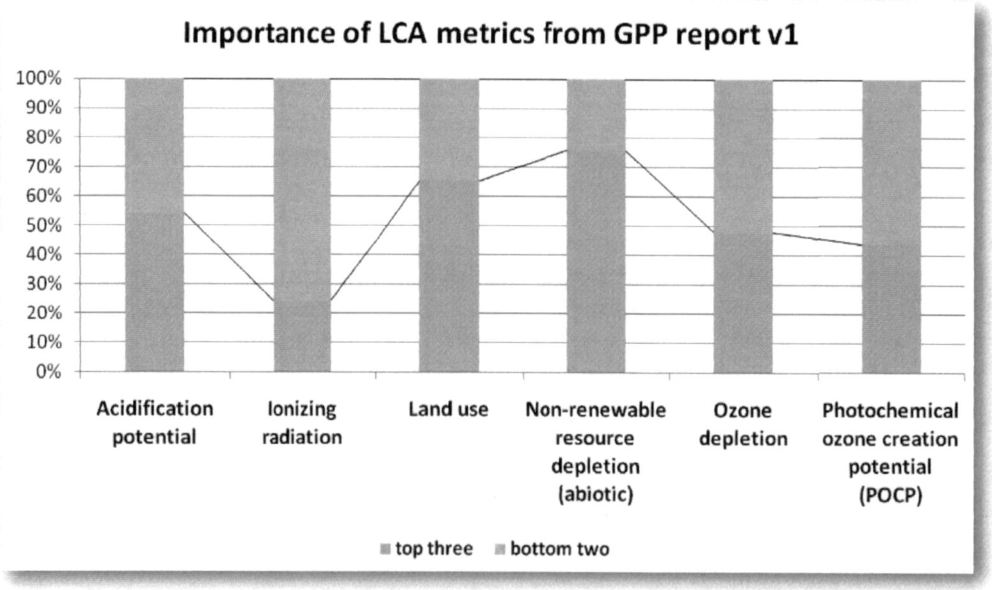

SUSTAINABLE PACKAGING COALITION®

a project of GreenBlue

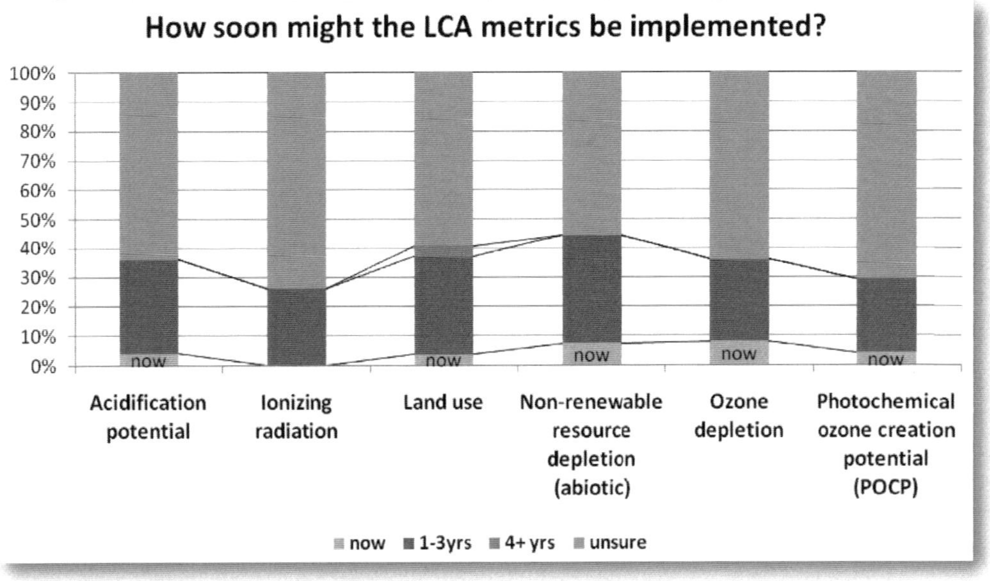

LCI DATA BOTTLENECK

SUSTAINABLE PACKAGING COALITION®
a project of GreenBlue
© 2011

Public Data Sources and Issues

Basic Fact

Anyone using publically available data are relying on the same sources of LCI.

- ecoinvent
  - o European focus
  - o System and unit-process level (~4000 entries)
  - o QC-ed and vetted
- U.S. LCI
  - o U.S. focus
  - o Unit-process level data (<300 entries)
  - o Not well QC-ed or vetted

SUSTAINABLE PACKAGING COALITION®

a project of GreenBlue

To Achieve Sustainability Priorities

- Benchmark packaging portfolio
  - o Start measuring key indicators that apply to your business
- Track changes
- Communicate with supply chain partners
- Share LCI data

To do this:

- One need not wait for the perfect tool
- Be consistent with methodology being used
- Understand the limitations of the data and LCA assessment
- Appreciate that LCA is only one tool to have in the tool box
- Start with a clear set of corporate sustainability priorities

SUSTAINABLE PACKAGING COALITION®

a project of GreenBlue

© 2011

Online Packaging Assessment

COMPASS Features

- Compare up to 4 packaging scenarios simultaneously
- Assess impact categories based on common functional units
- View each component's impact in relation to the package
- Assess consumption and emission metrics
- Assess solid waste profile of design
- Include distribution impacts
- Easy to use web-based secure application
- Consistent data modeling for packaging applications

SUSTAINABLE PACKAGING COALITION®

a project of GreenBlue

COMPASS®
COMPARATIVE PACKAGING ASSESSMENT

https://www.design-compass.org/

Free trial available online

MINAL T. MISTRY
PROJECT MANAGER
minal.mistry@greenblue.org
www.sustainablepackaging.org
https://www.design-compass.org

SUSTAINABLE PACKAGING COALITION®

a project of GreenBlue

- Introduction
- Packaging LCA Overview
- Packaging LCA at Nestlé
- PIQET as a Screening LCA Tool at Nestlé
- Comprehensive Packaging LCA Example for Nestlé

LIFE CYCLE THINKING FOR PACKAGING

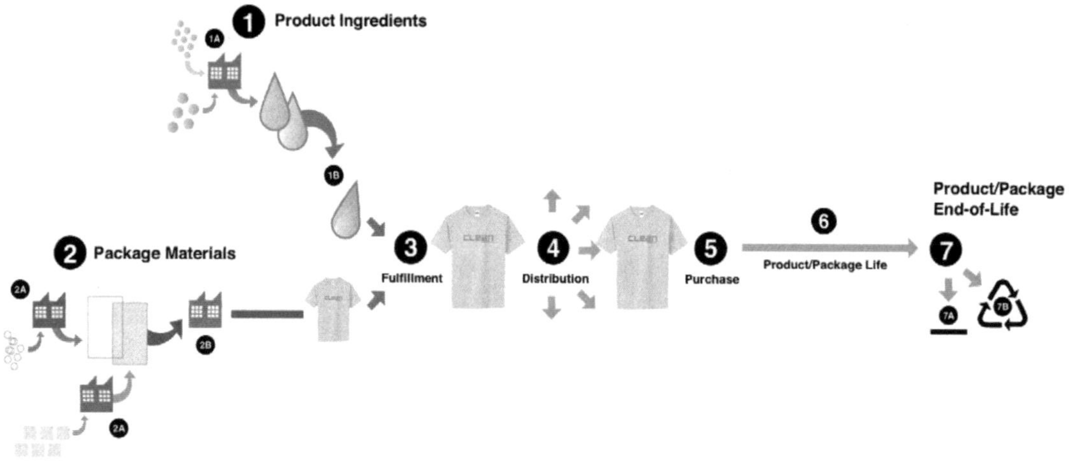

SAMPLE LIFE CYCLE - FRAMEWORK AND SYSTEM BOUNDARIES

CRADLE TO GRAVE

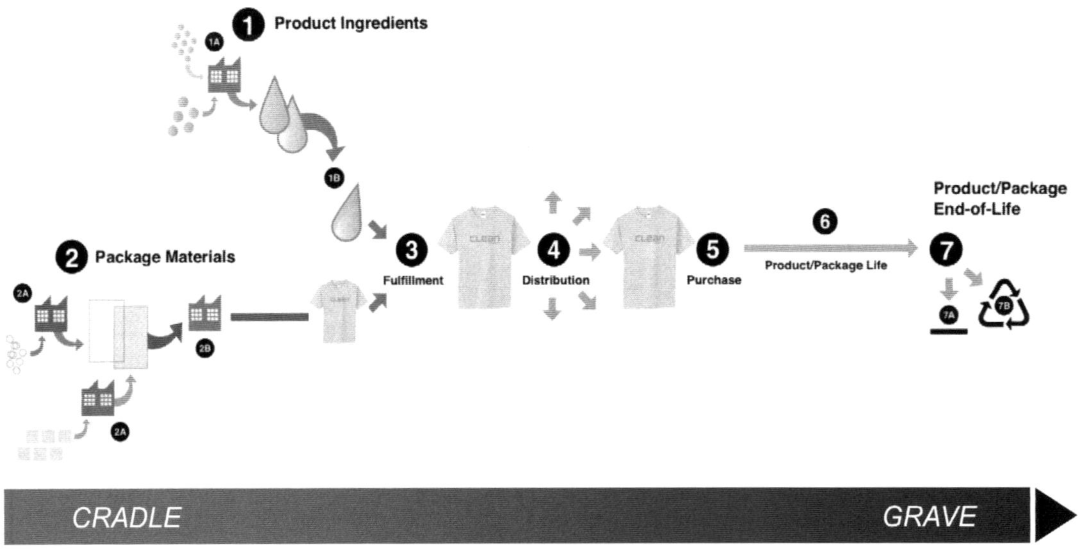

SAMPLE LIFE CYCLE - SYSTEM BOUNDARIES

CRADLE TO GATE

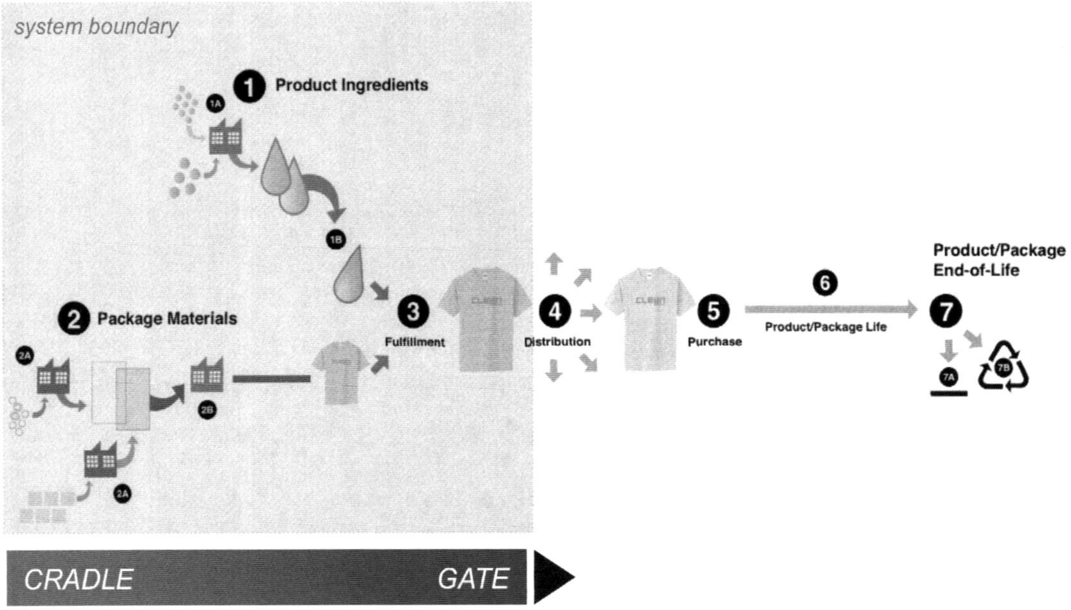

SAMPLE LIFE CYCLE - SYSTEM BOUNDARIES

GATE TO GATE

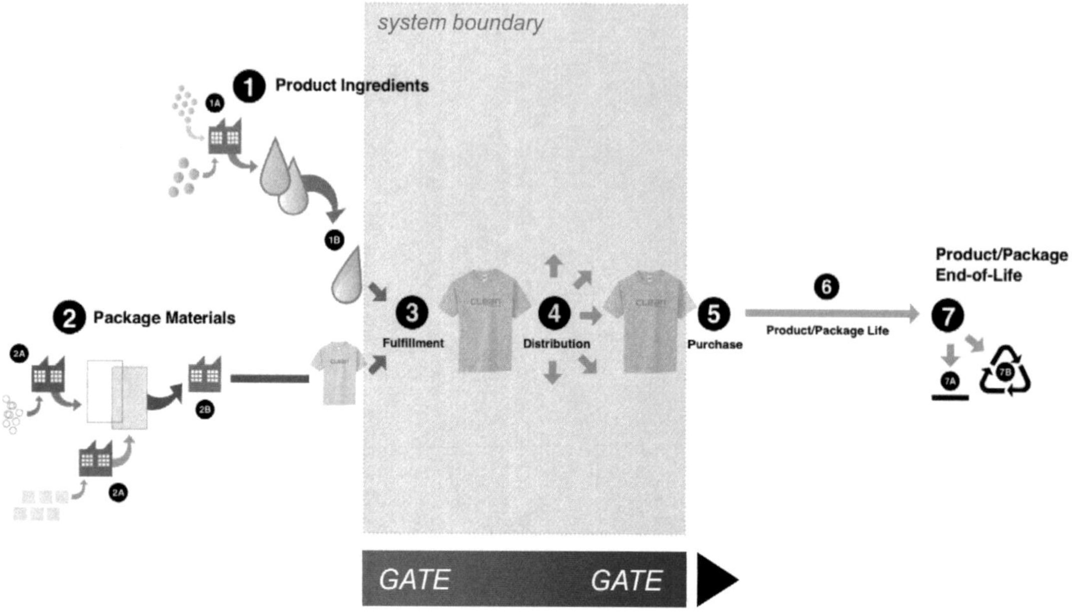

MAJOR ELEMENTS

LIFE CYCLE ASSESSMENTS

- Define **goal** of study and **functional unit** (FU) of product or service (GOAL)
- **System map** of delivery of FU (SCOPE)

- **Activity data** (e.g., bill of materials for energy, water, and materials)
- **Inventory** of resource consumption and pollutants
- **Impact** modeling for fates, eco or human toxicity to measure environmental damage
- **Interpretation** (Uncertainty analysis, sensitivity analysis, etc.)

Bottom line value of LCA is to raise **awareness of environmental impacts** of products or processes that might otherwise be overlooked.

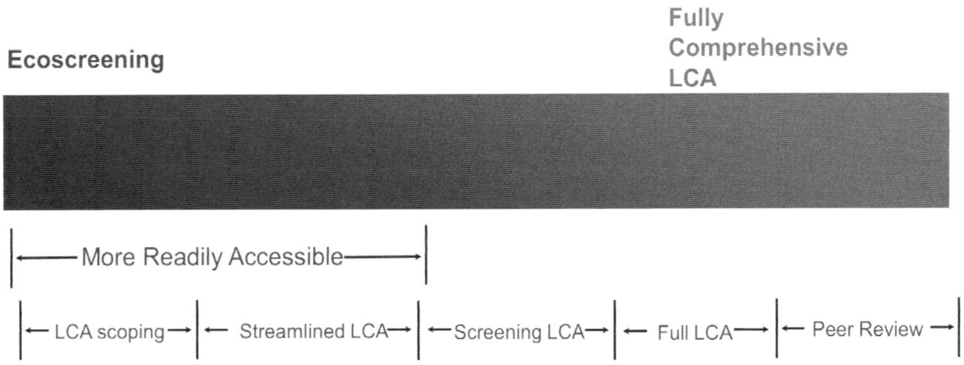

MAJOR ELEMENTS

HOW IS THE STUDY **GOAL** RELATED TO THE **SCOPE?**

- A key concern with LCA's is the **time and resources** required
- With **refinement of the goal**, the scope can often be narrowed
- **Limited scope** of just one aspect of a product system might be appropriate for a specific comparison or **broader scope** might be needed for reaching into system design elements

LCA based decision support: Maximised influence at minimum cost

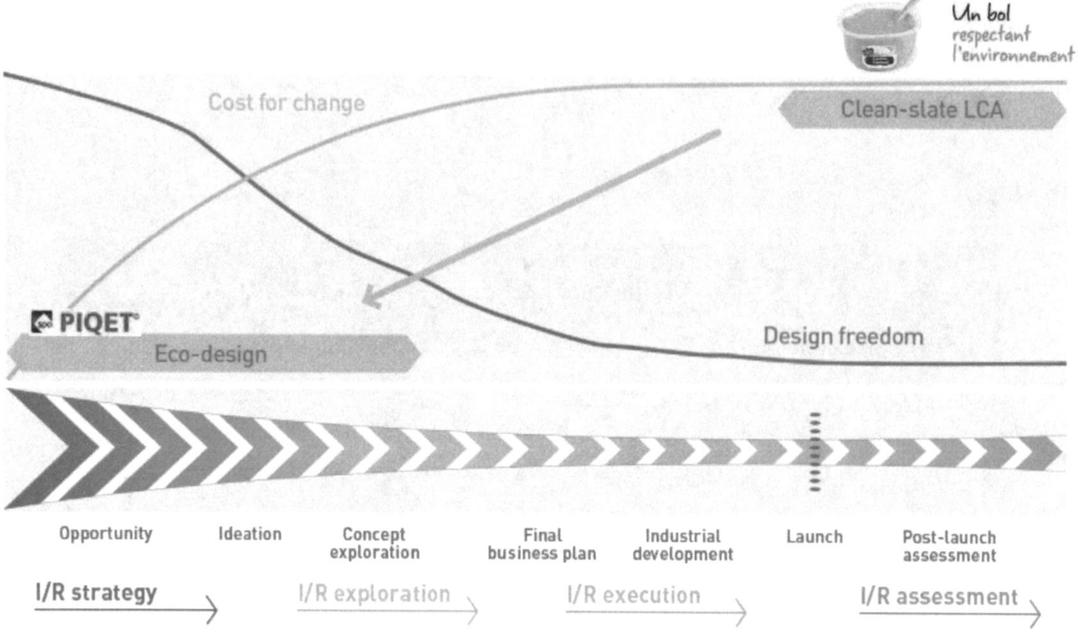

Functional Units in Packaging

- Protection of product
- Theft deterrence
- On-shelf performance
- Supply chain efficiencies

The Nestlé Story

- 456 factories in 84 countries
- 280,000 employees
- Sales of 107.6 billion in 2009
- 135,000 different products (more than 1 billion products sold every day)

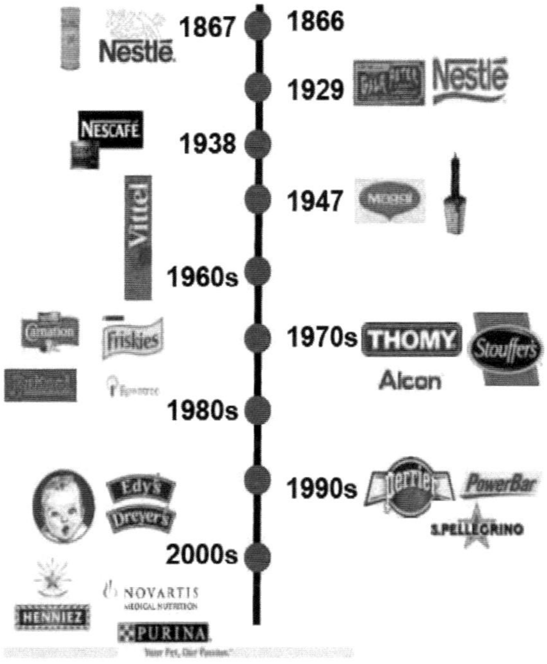

Nestlé Environmental Sustainability Policy

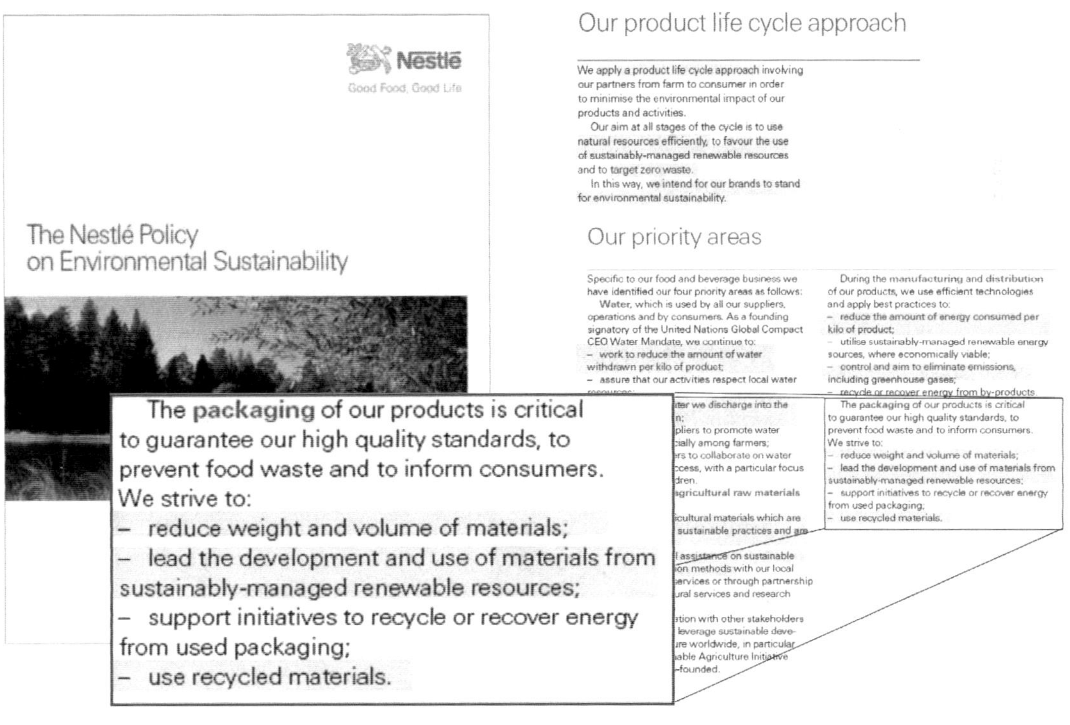
Nestlé
Good Food, Good Life

The Nestlé Policy on Environmental Sustainability

Our product life cycle approach

We apply a product life cycle approach involving our partners from farm to consumer in order to minimise the environmental impact of our products and activities.

Our aim at all stages of the cycle is to use natural resources efficiently, to favour the use of sustainably-managed renewable resources and to target zero waste.

In this way, we intend for our brands to stand for environmental sustainability.

Our priority areas

Specific to our food and beverage business we have identified our four priority areas as follows:

Water, which is used by all our suppliers, operations and by consumers. As a founding signatory of the United Nations Global Compact CEO Water Mandate, we continue to:
- work to reduce the amount of water withdrawn per kilo of product;
- assure that our activities respect local water resources;

During the manufacturing and distribution of our products, we use efficient technologies and apply best practices to:
- reduce the amount of energy consumed per kilo of product;
- utilise sustainably-managed renewable energy sources, where economically viable;
- control and aim to eliminate emissions, including greenhouse gases;
- recycle or recover energy from by-products.

The packaging of our products is critical to guarantee our high quality standards, to prevent food waste and to inform consumers. We strive to:
- reduce weight and volume of materials;
- lead the development and use of materials from sustainably-managed renewable resources;
- support initiatives to recycle or recover energy from used packaging;
- use recycled materials.

The packaging of our products is critical to guarantee our high quality standards, to prevent food waste and to inform consumers. We strive to:
- reduce weight and volume of materials;
- lead the development and use of materials from sustainably-managed renewable resources;
- support initiatives to recycle or recover energy from used packaging;
- use recycled materials.

PIQET©

Packaging Impact Quick Evaluation Tool

- Web-based interface to SimaPro LCA tool
- Adapted to FMCG packaging development with
  - User interface adapted to use by non-experts
  - Short turn around time (approximately 30 minutes with data available for experienced user)
- Pre-defined functional unit: a kilogram of packaged product delivered to a customer
- Packaging life cycle improvement analysis
  - Scenario comparison (trend-based decision support)
  - Permit evaluation of trade-offs between impact categories
  - Database for materials & processes for various regions

PIQET Spider Graphs

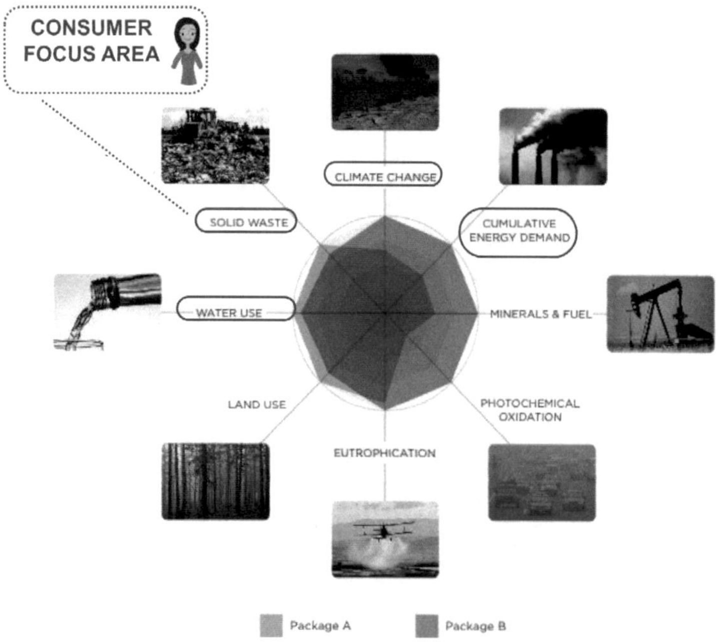

Graphs show **comparative environmental impacts per kg delivered** from 2-3 different ways of packaging products.

Provides an overview of trade-offs between **8 different environmental impacts.**

Used to **assist in packaging design and material selection decisions**, but data will always need to be interpreted carefully and weighed with other requirements.

Differences **should generally be at least 20%** to determine a preference for a given alternative.

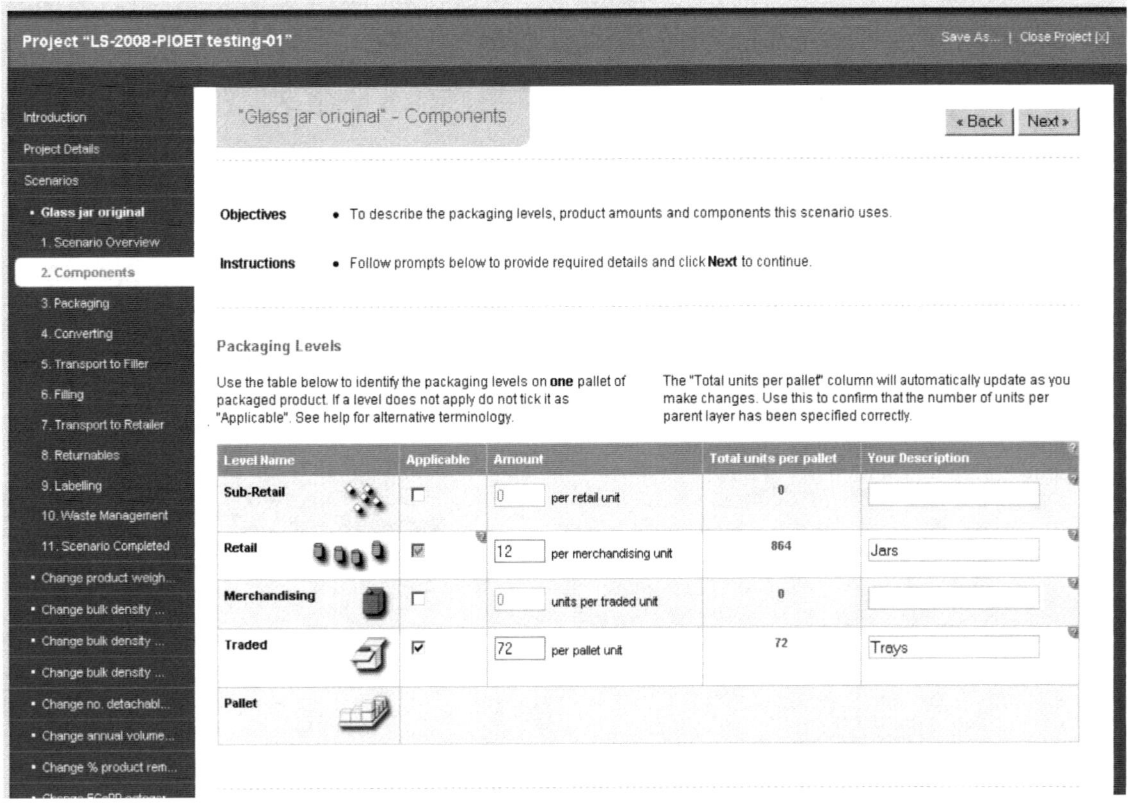

PIQET in Practice: Illustration of Life Cycle Phases

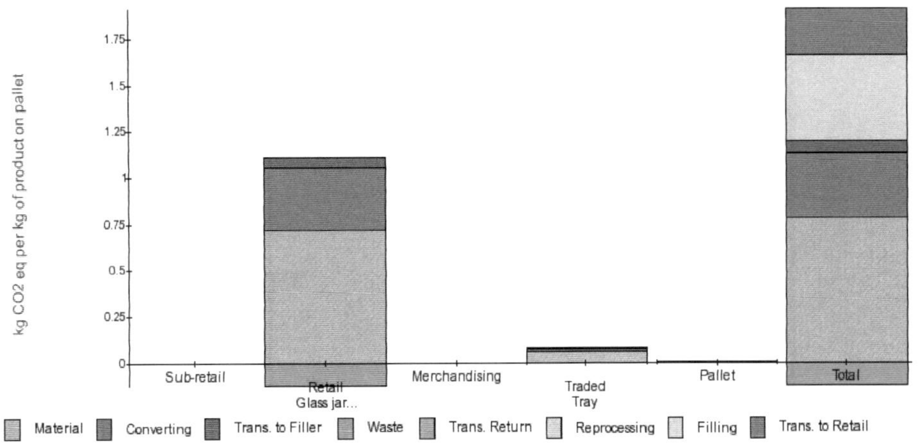

PIQET in Practice:
Illustration of Life Sub-Component Analysis

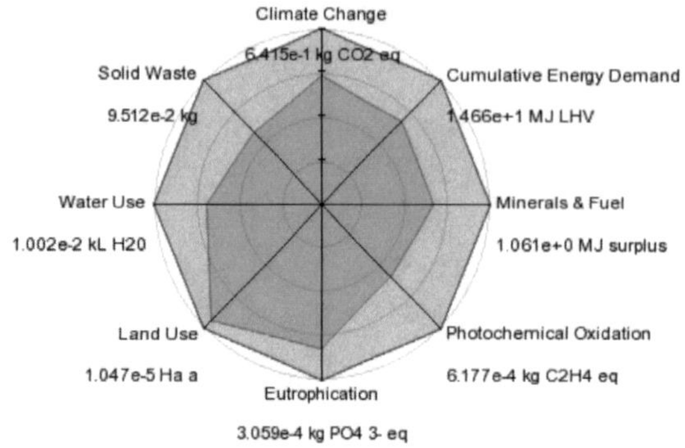

Except for land use (no deviation in natural/fiber materials) there are large differences in the other environmental impact indicators.

Spidergraph suggests that there are redesigns should focus on reducing the size of the lid.

With Lid/Fitment

No Lid/Fitment

LCA of baby food packaging

Packaging systems & scenarios

Glass Jar components | **Plastic Pot components**

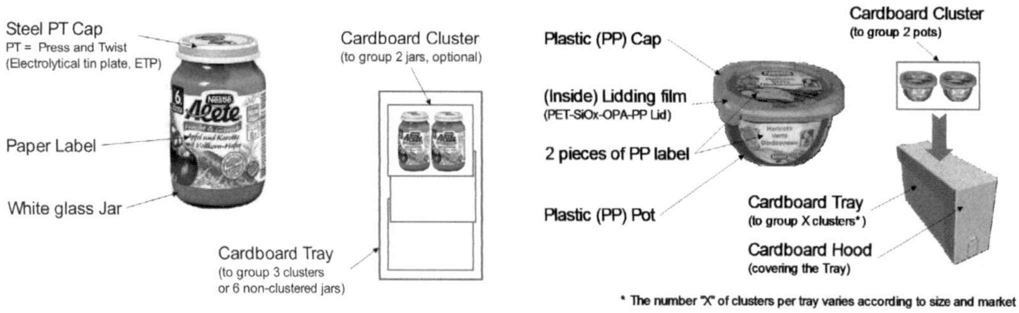

Retorting → Aseptic form-fill-seal

S. Humbert et al.,"Life cycle assessment of two baby food packaging alternatives: glass jars vs. plastic pots", *International Journal of Life Cycle Assessment* Volume 14, Number 2 / March, 2009

From PIQET to Full LCA

- Details on filling energy and chemical use
- Detailed logistical info on truck loadings and modes (weighted averages)
- Scrap waste at supplier and filling factory
- Details for all packaging levels and components (1% cut-off rule)
- Lots of primary data
- Full system expansion for End-of-Life
- Extensive sensitivity and uncertainty analysis

Full LCA Answers: Tradeoffs & Value Judgements

- 14% Less Primary Energy
- 28% Less Global Warming
- 31% Less Respiratory Inorganics
- 28% Less terrestrial acidification/nutrification
- **No** measure for **water**
- **No** measure for **consumer solid waste**
- Ionizing radiation worse for France

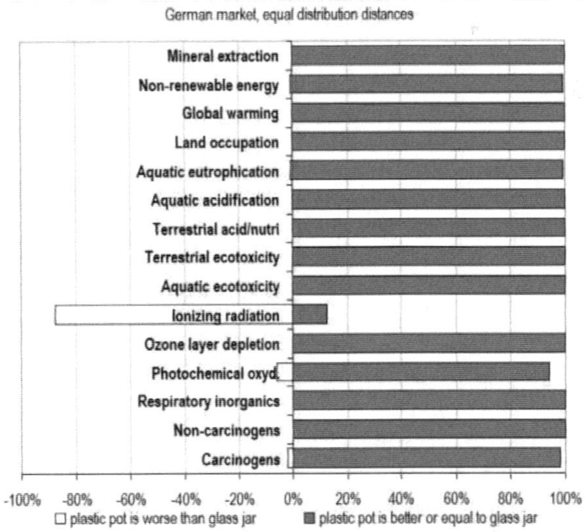

Summary of Nestlé's Packaging LCA Process

- Packaging innovation & renovation process
- Systematic use of streamlined LCA-based decision support
- Life Cycle Assessment screening studies

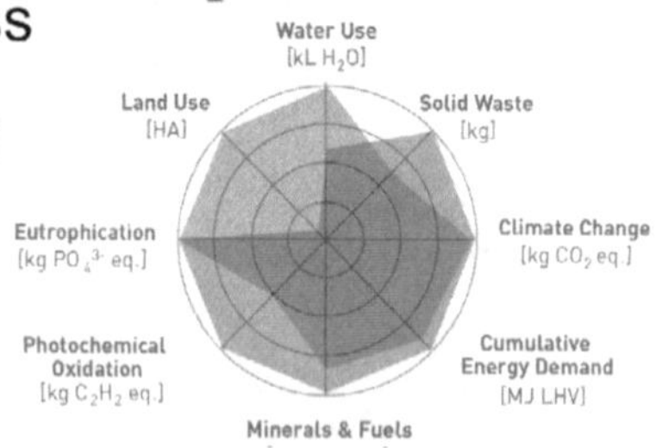

- Communication of claims
- Full LCA studies according to ISO 14040 &140444
- Data quality and representativeness
- Uncertainty & sensitivity analysis
- Peer review

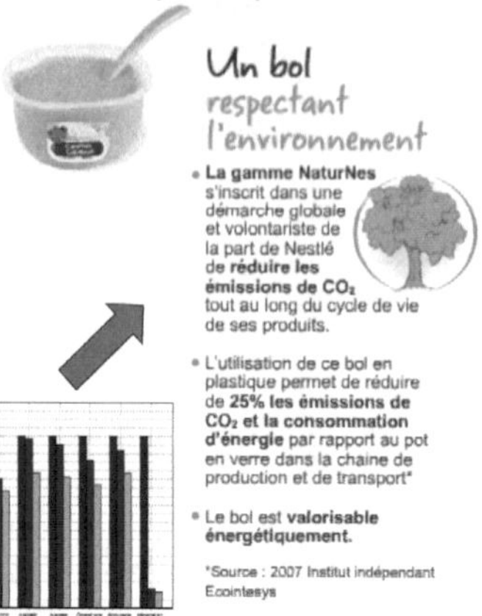

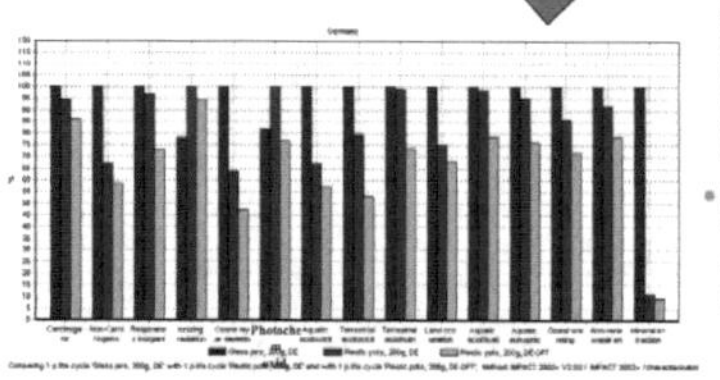

DISCUSSION & QUESTIONS

For More Info on Packaging LCA

Robert Vos, Ph.D.

Research Director

Clean Agency

rvos@cleanagency.com

For More Info on PIQET:

Victor Barichello

Marketing Manager

Sustainable Packaging Alliance

victor@sustainablepack.org

Evaluating the social, economic and environmental consequences that flow from your activities.

EarthShift
Building pathways to sustainability

Sustainable Packaging Symposium
Chicago, IL

Laurel McEwen
laurel@earthshift.com

830 Taft Road • Huntington, VT 05462 • Phone (802) 434-3326 Fax (802) 329-2214 • www.earthshift.com

Critical aspects for success of a simplified tool

1. Support the creative process
   (don't cage me in)

2. Simplify....but integrate professional LCA knowledge
   (a little history)

3. Make environmental decisions early
   (environmental leverage is at the beginning)

4. Apply LCA methodology consistently/accurately
   (easy to mess it up)

How?

Leverage the iterative nature of design and LCA
(lets blend them)

EarthShift

1. Creative Process....investigate

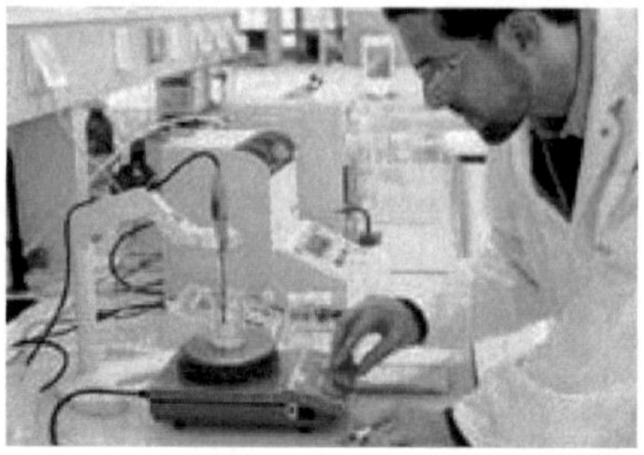

knowing the details we solved
the problems and created new ideas

EarthShift

1. Creative Process....requires answering the "Why?"

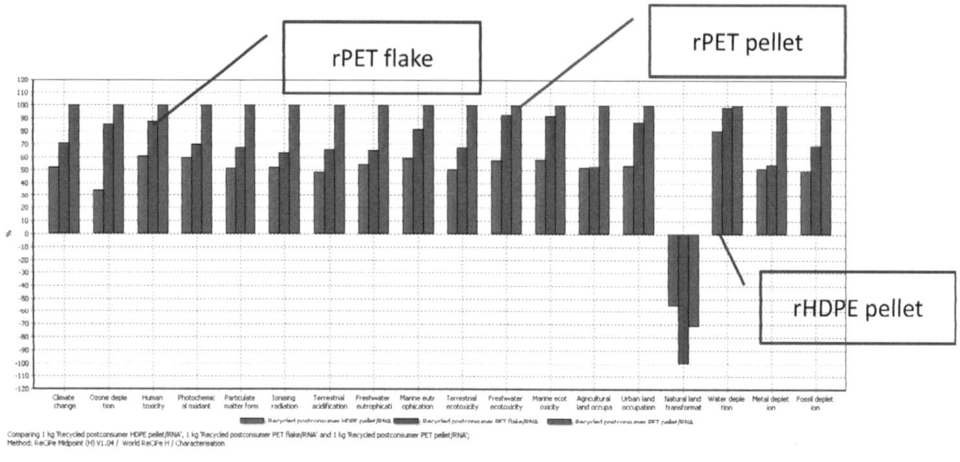

If you don't tell a designer
why something doesn't work......
they can not work creatively

EarthShift

1. Creative Process....brain drain

How can I create,
when I don't know the rules?

EarthShift

2. The Integration of LCA into Companies
......a little history

➡ 1st
Hire external consultancy

➡ 2nd
Form internal power users

A typical LCA:
costs: $25,000 - $50,000
takes: 3 – 1 yr months

...and they have learned that to affect real change they have to integrate LCA into the design process

EarthShift

3. Environmental leverage is at the beginning

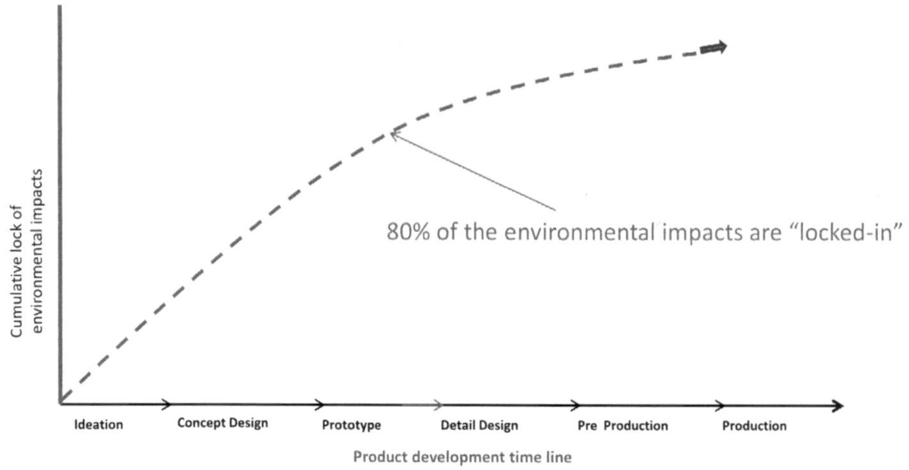

...where design freedom is high, and the cost of change is low

EarthShift

LCA eze

Functional Unit

PDF*m2*yr

MJ primary

Cut Off

USEtox

Reference Flow

DALY

Boundary Conditions

Economic Allocation

Pedigree Matrix

Input/Output

Avoided Burden

Open-Loop

Allocation, exergy

ReCiPe

Closed-Loop

EarthShift

Ideal Simplified Tool Requirements

1. Include early in design phase
2. Non LCA expert can use
3. Affordable
4. Can answer "Why?"

and there is one more thing………

EarthShift

4. Methodological Choices
...must be applied consistently and accurately

- Primary LCI data collected
- Secondary LCI identified/modified
- Recycling model chosen
- Allocation method defined
- Impact Assessment Method selected
- Normalization reference selected
- Weighting defined

....so all the product development teams are making consistent choices

EarthShift

4. Methodological Choices
...must be applied consistently and accurately

Back to the drawing board

Leverage is lost

EarthShift

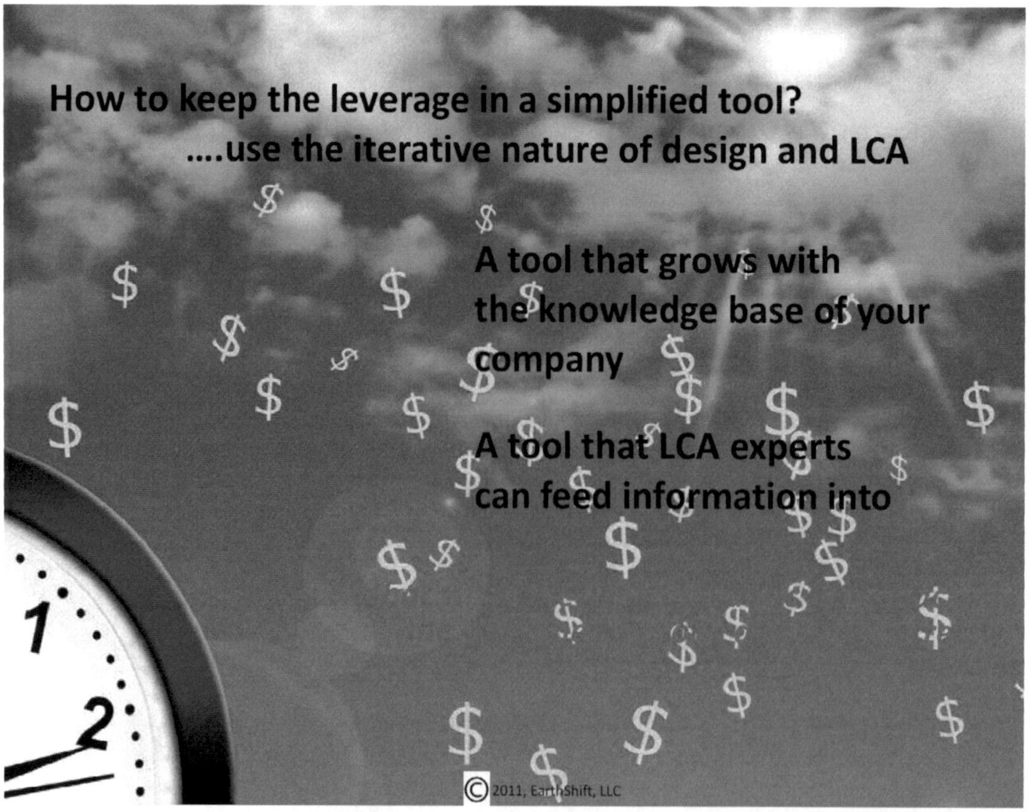

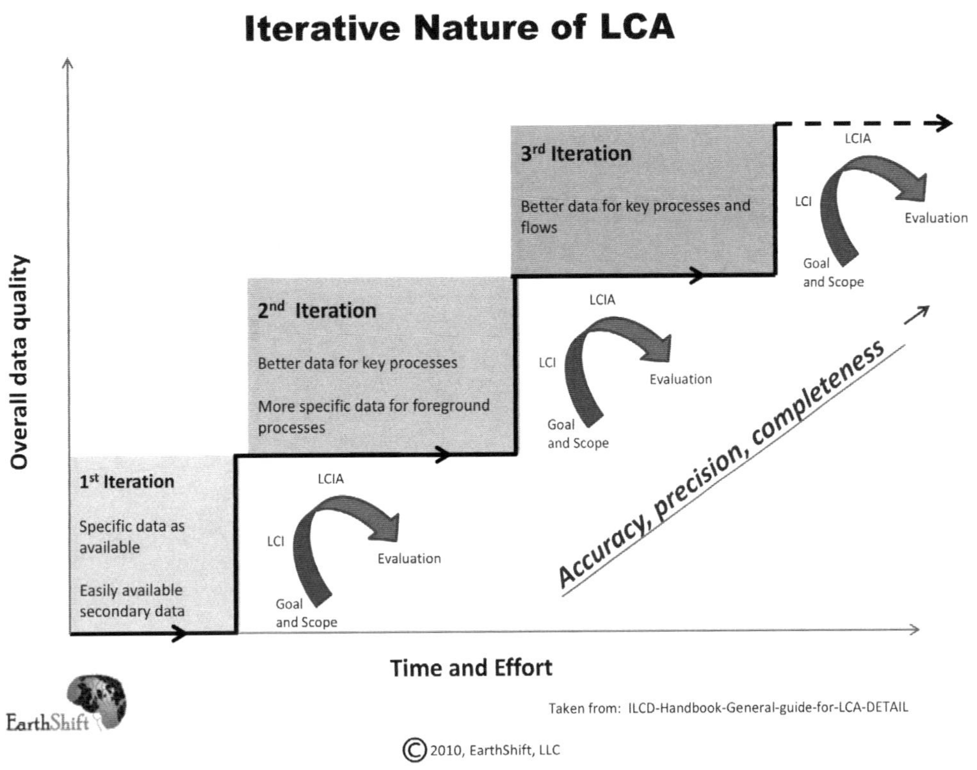
Iterative Nature of LCA
Overall data quality
1st Iteration
Specific data as available
Easily available secondary data
2nd Iteration
Better data for key processes
More specific data for foreground processes
3rd Iteration
Better data for key processes and flows
LCIA
LCI
Evaluation
Goal and Scope
Accuracy, precision, completeness
Time and Effort
EarthShift
Taken from: ILCD-Handbook-General-guide-for-LCA-DETAIL
©2010, EarthShift, LLC

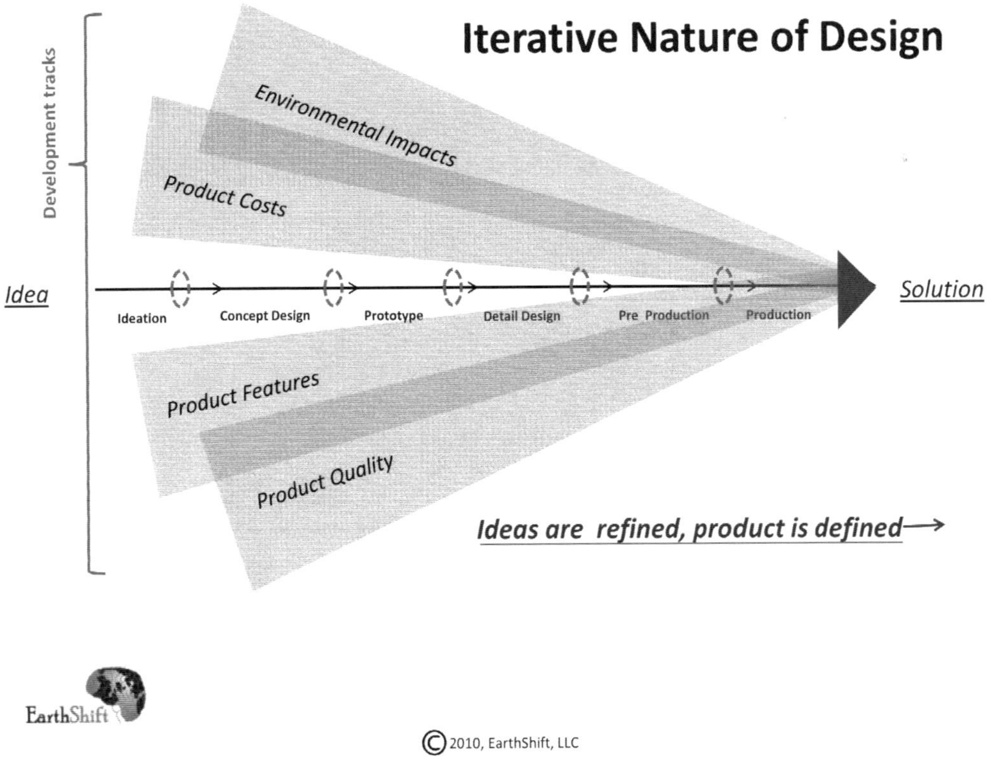
Iterative Nature of Design
Development tracks
Environmental Impacts
Product Costs
Idea
Ideation
Concept Design
Prototype
Detail Design
Pre Production
Production
Solution
Product Features
Product Quality
Ideas are refined, product is defined
EarthShift
©2010, EarthShift, LLC

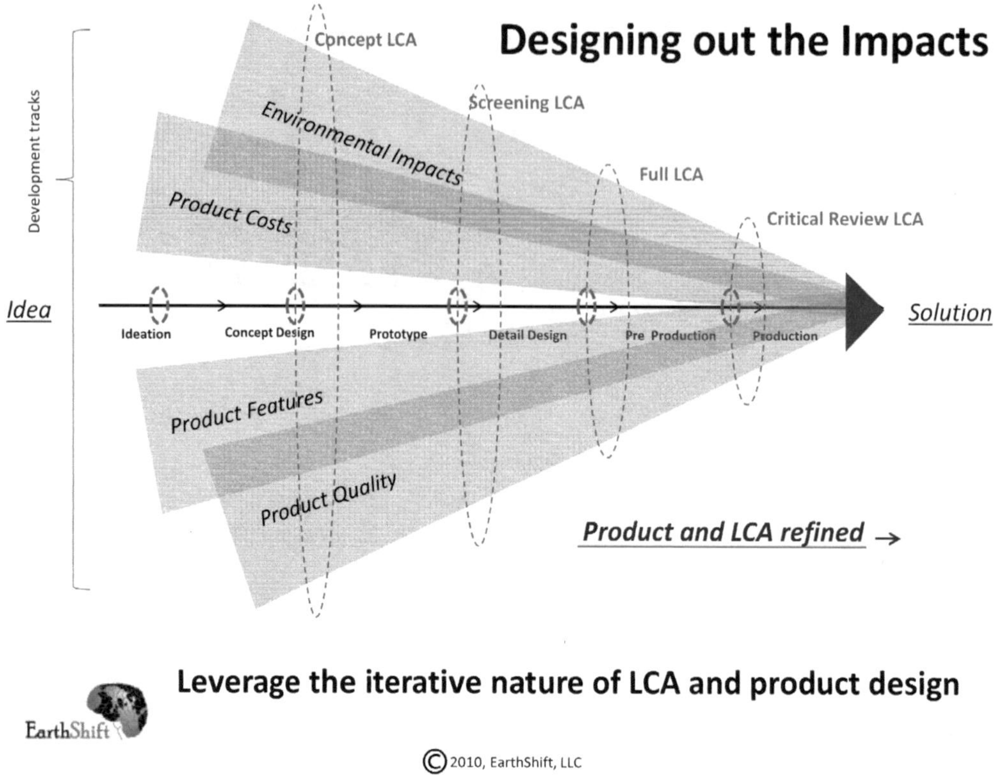
Designing out the Impacts
Concept LCA
Screening LCA
Full LCA
Critical Review LCA
Development tracks
Environmental Impacts
Product Costs
Product Features
Product Quality
Idea
Ideation
Concept Design
Prototype
Detail Design
Pre Production
Production
Solution
Product and LCA refined
Leverage the iterative nature of LCA and product design
EarthShift
©2010, EarthShift, LLC

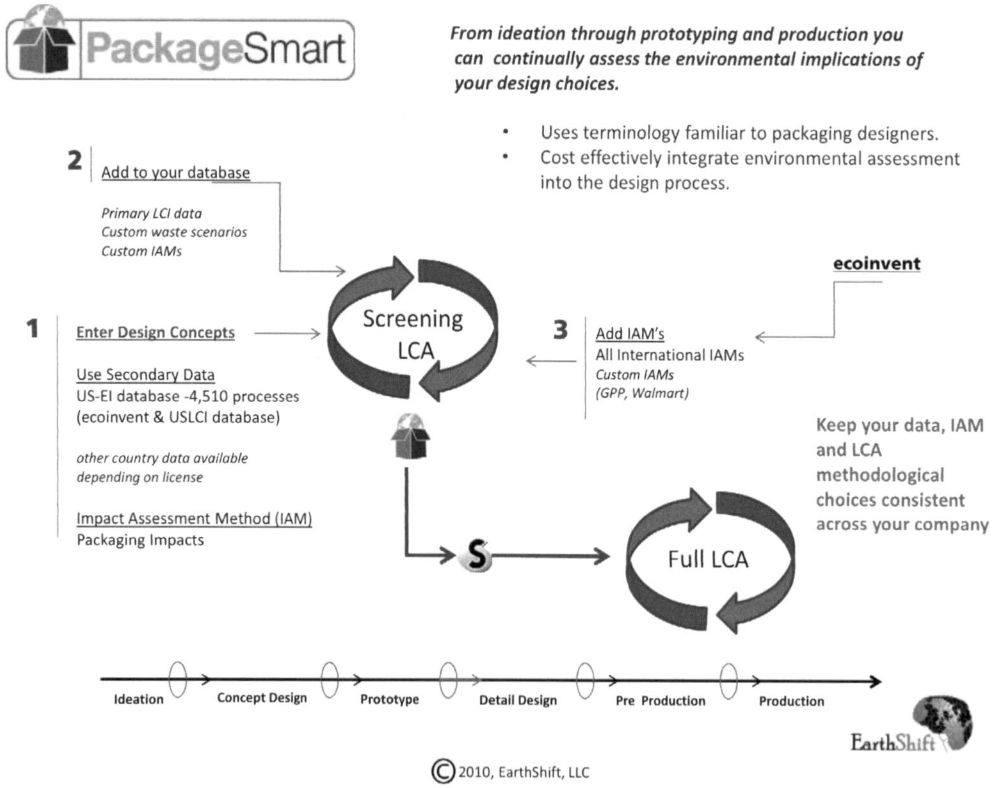
PackageSmart
From ideation through prototyping and production you can continually assess the environmental implications of your design choices.
• Uses terminology familiar to packaging designers.
• Cost effectively integrate environmental assessment into the design process.
2 Add to your database
Primary LCI data
Custom waste scenarios
Custom IAMs
1 Enter Design Concepts
Use Secondary Data
US-EI database -4,510 processes (ecoinvent & USLCI database)
other country data available depending on license
Impact Assessment Method (IAM)
Packaging Impacts
Screening LCA
3 Add IAM's
All International IAMs
Custom IAMs
(GPP, Walmart)
ecoinvent
Keep your data, IAM and LCA methodological choices consistent across your company
Full LCA
Ideation
Concept Design
Prototype
Detail Design
Pre Production
Production
EarthShift
©2010, EarthShift, LLC

PackageSmart

1 Select and name approved processes

Manager
Users
Viewers

Designing out the Impacts

Managers can give external consults access to user groups
Managers select approved processes
Managers select approved Impact Assessment Methods

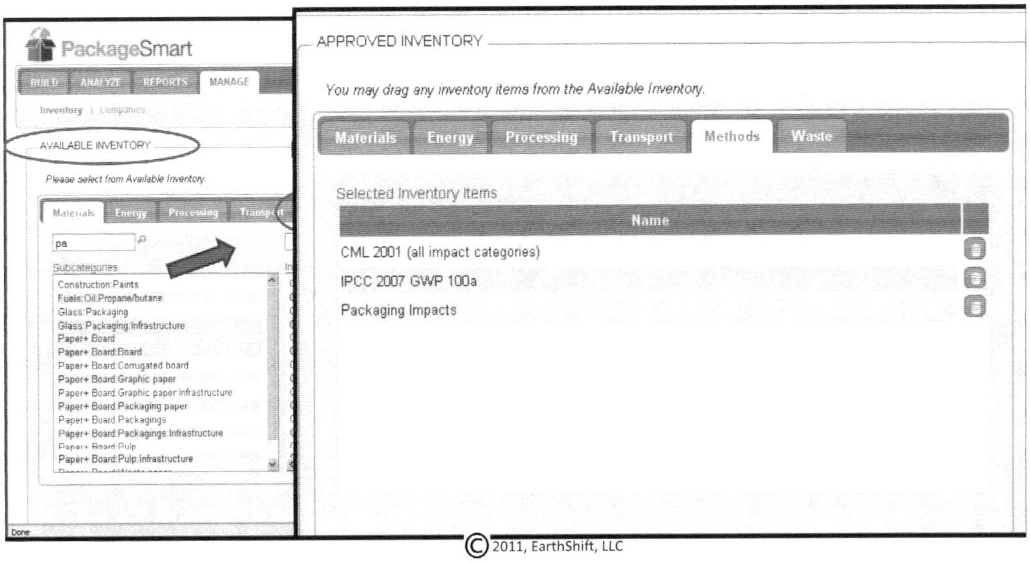

PackageSmart

2 Create a Project

Manager
Users
Viewers

Designing out the Impacts

Select waste scenario defaults
Select impact assessment method
Show cube efficiency, primary, secondary, tertiary

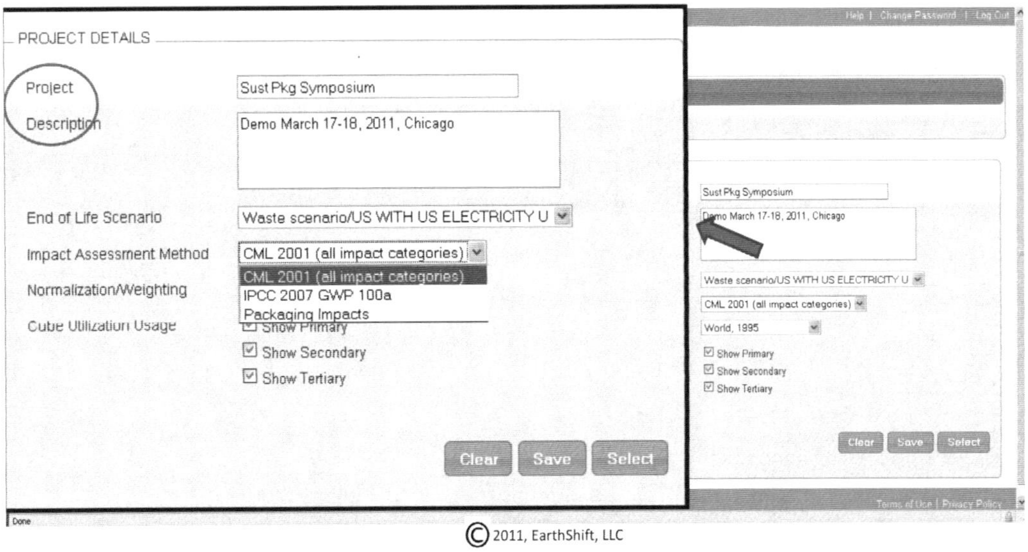

PackageSmart

Designing out the Impacts

3 Build Subassemblies

- **Manager**
- **Users**
- Viewers

Select materials, enter amount
Select processing, enter amount
Select transport, enter distance
Select energy, enter amount

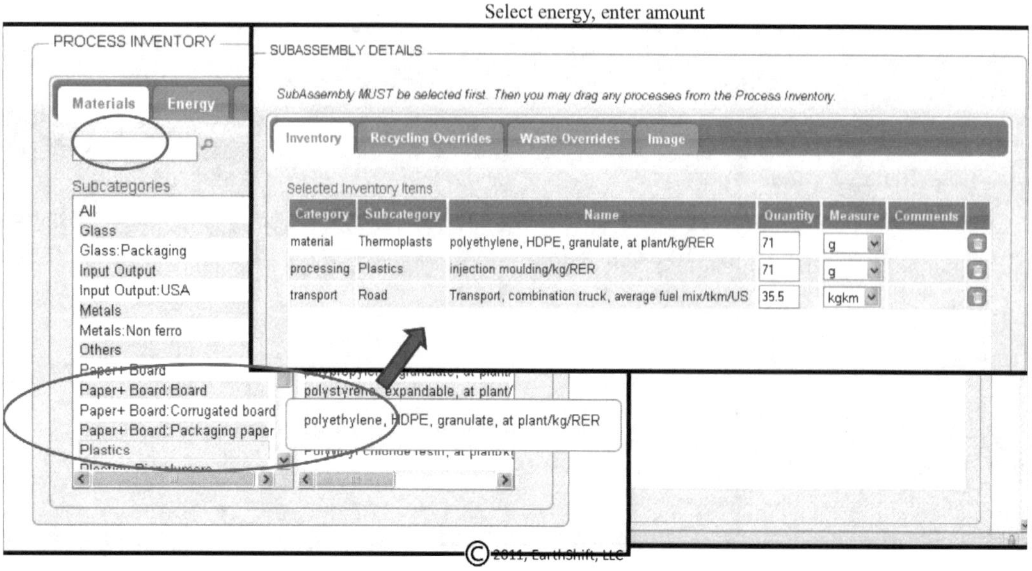

PackageSmart

Designing out the Impacts

4 Customize End of Life

- **Manager**
- **Users**
- Viewers

Customize recycling rates
Customize end of life incineration/landfill rates

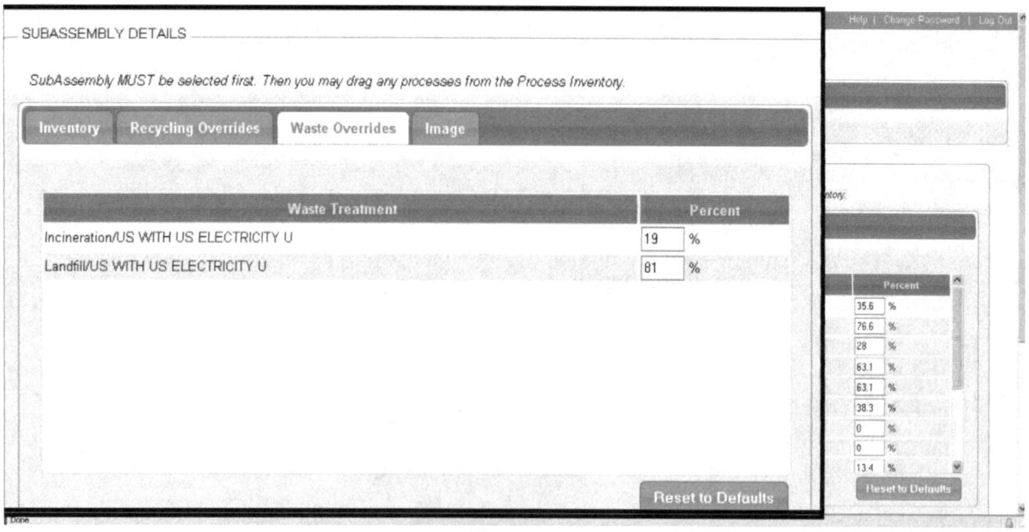

PackageSmart

Designing out the Impacts

5 **Add Image**

Manager
Users
Viewers

Drag or browse to an image file

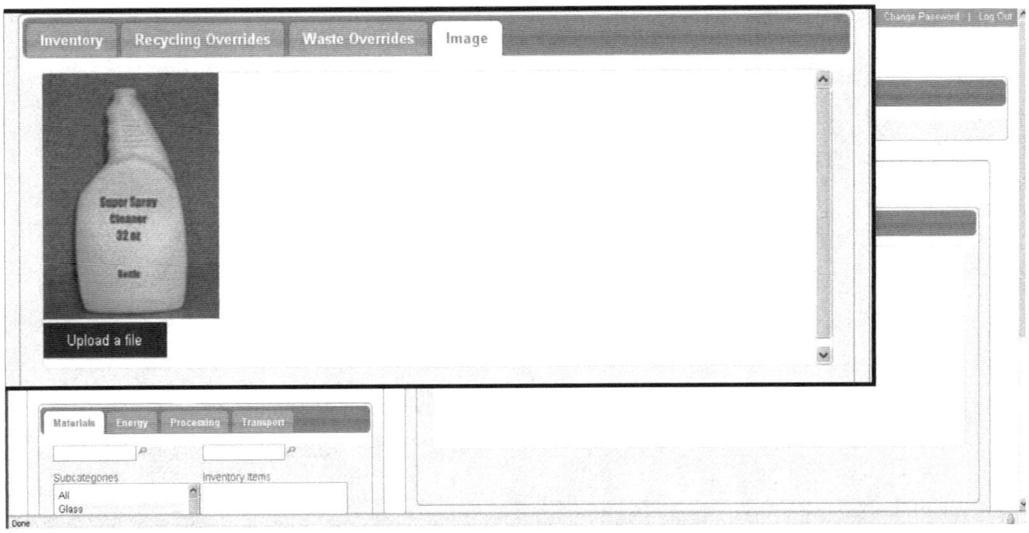

PackageSmart

Designing out the Impacts

6 **Build Assemblies**

Manager
Users
Viewers

Add sub-assemblies into an assembly, enter quantity
Define the number of uses

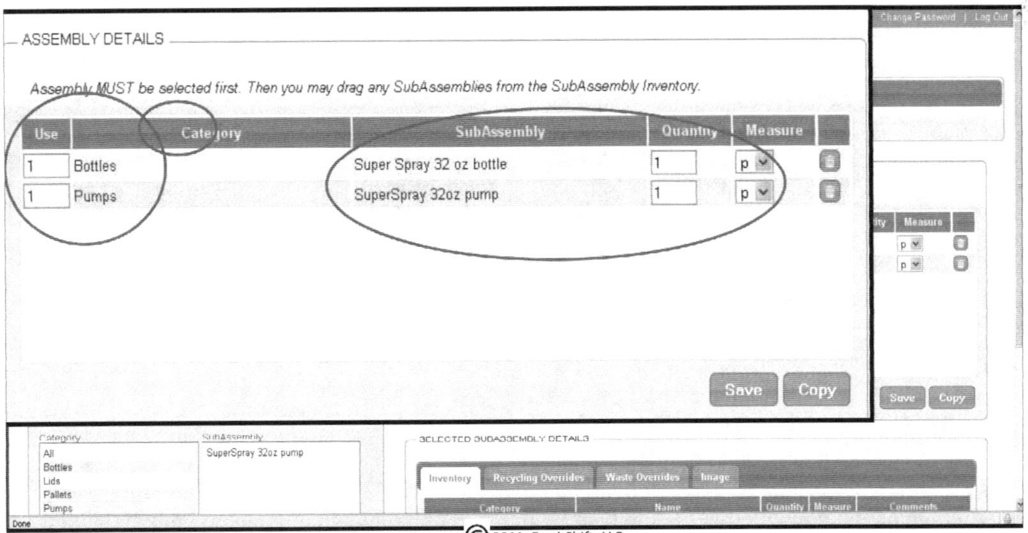

7 Build Package

Manager
Users
Viewers

Designing out the Impacts

Add primary, secondary and tertiary packaging
Enter number of uses
Add cube utilization details if wanted
Add distribution transportation

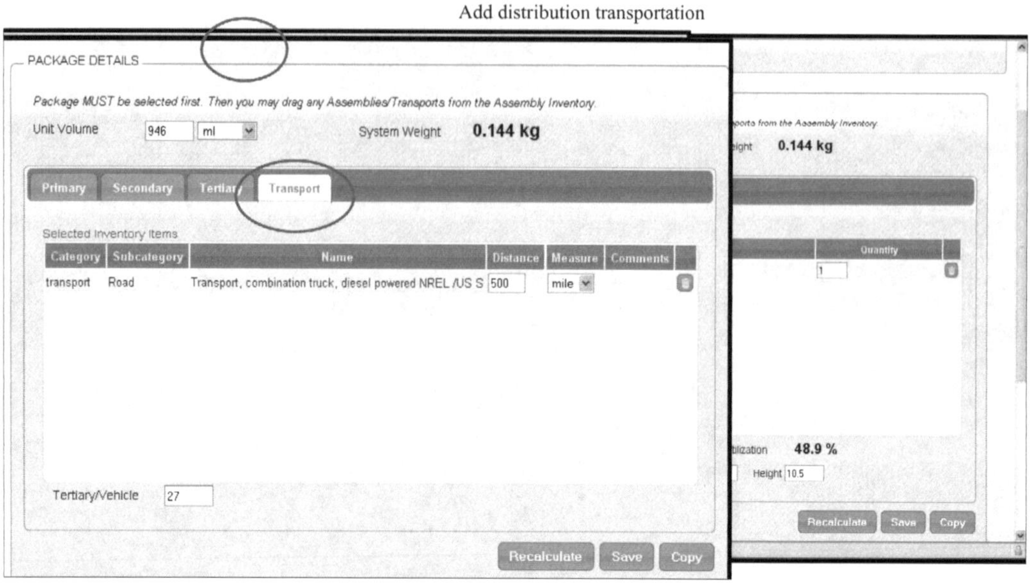

8 Run Analysis

Designing out the Impacts

t /unit of comparison
ment method default
rmalized/weighted scores

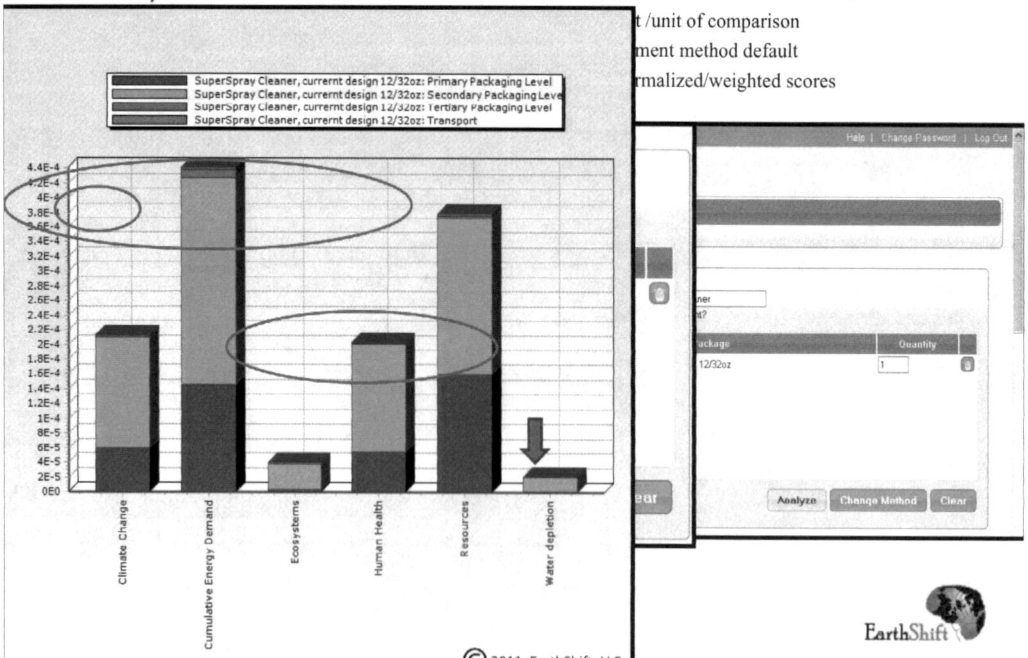

EarthShift

Designing out the Impacts

8 Create Report

- Manager
- Users
- Viewers

Managers can create custom report templates

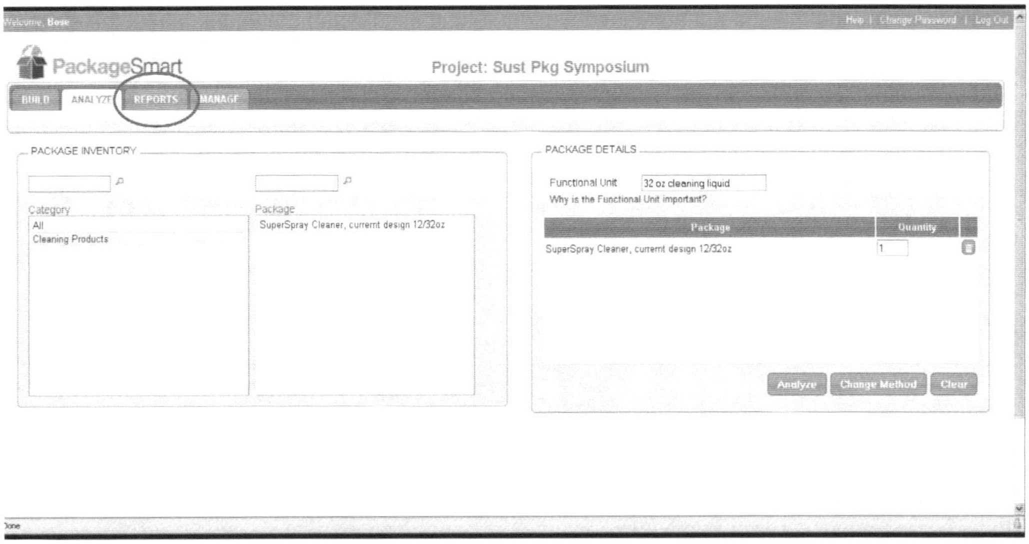

PackageSmart enables R&D designers to realize complete and detailed environmental evaluations, as well as create quick decision-making studies during the design process.

EarthShift

830 Taft Road • Huntington, VT 05462 • Phone (802) 434-3326 Fax (802) 329-2214 • www.earthshift.com

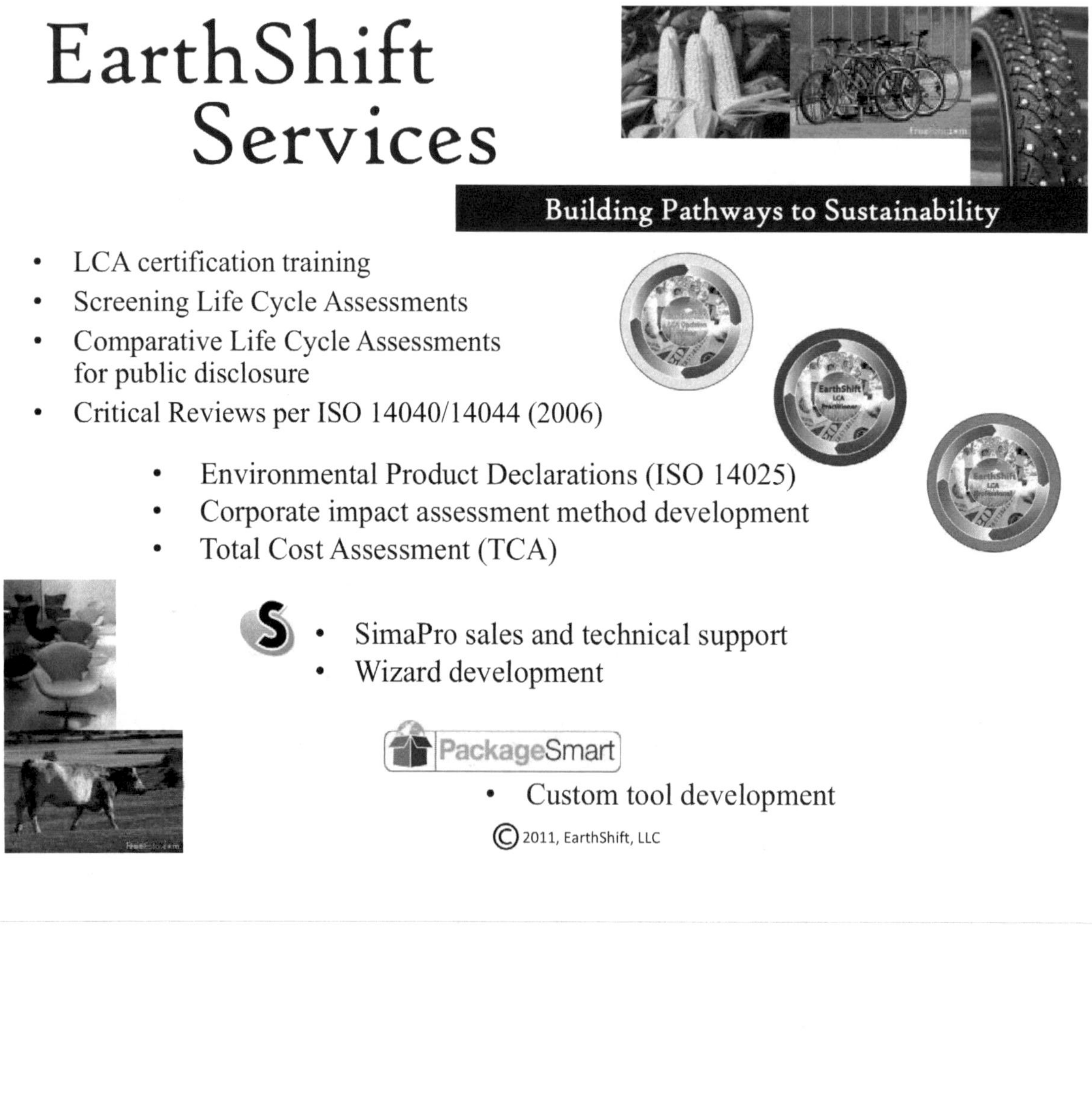
EarthShift
Services
Building Pathways to Sustainability
• LCA certification training
• Screening Life Cycle Assessments
• Comparative Life Cycle Assessments for public disclosure
• Critical Reviews per ISO 14040/14044 (2006)
• Environmental Product Declarations (ISO 14025)
• Corporate impact assessment method development
• Total Cost Assessment (TCA)
• SimaPro sales and technical support
• Wizard development
PackageSmart
• Custom tool development
© 2011, EarthShift, LLC

Five Winds INTERNATIONAL

PE INTERNATIONAL
EXPERTS IN SUSTAINABILITY

Sustainable Packaging Symposium
Laura Flanigan
March 18, 2011

Understanding, Managing Tradeoffs and Creating Opportunities

Or: How I learned to stop worrying
and love multi attribute LCA

Elements of Change

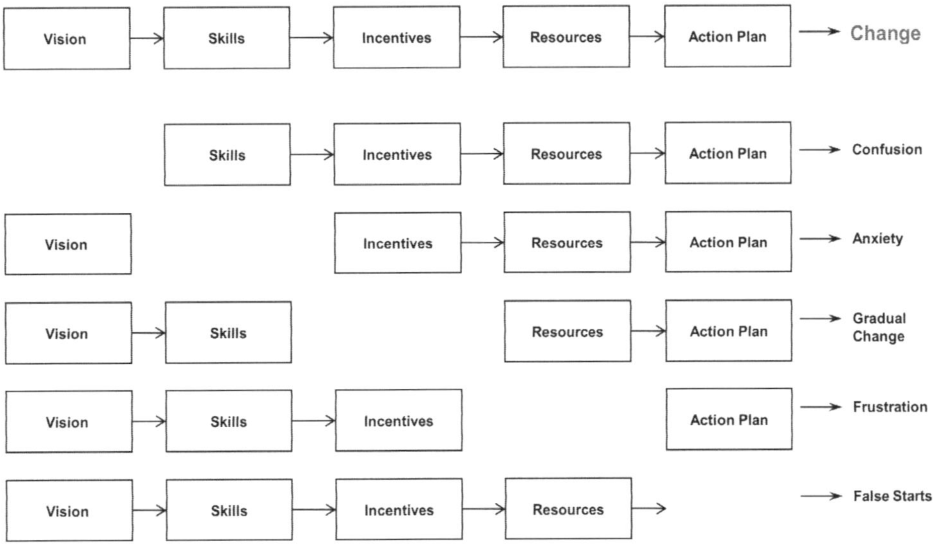

Source: George Manning, Kent Curtis: "The Art of Leadership" (2002)

Introduction to PE Five Winds

Five Winds International

PE INTERNATIONAL

Integrated solutions for product and corporate sustainability management

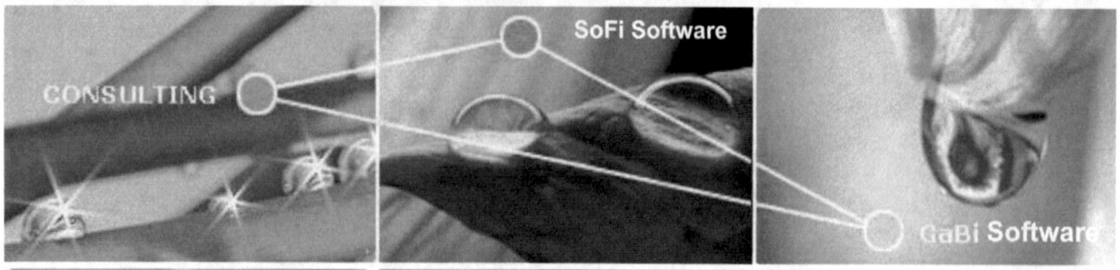

Strategy Solutions	Management Tools	Product Evaluations
• Sustainability Strategy • Carbon Management • Stakeholder Engagement • Design for Environment • Green Claims and Marketing	• Management Systems • Data Management Solution • GRI, CDP, UN Global Compact, ISO 14001, OHSAS 18000, etc. • Corporate Carbon Footprint	• Product Life Cycle Assessment • Product Carbon Footprint • Scenario analysis / decision support

Databases & Content

Abbreviated List of Clients and Partners

Five Winds International

PE INTERNATIONAL

Our customers succeed with our solutions.

Electronics	NOKIA Connecting People, LG, DELL, SONY, CISCO, SIEMENS, intel
Materials	Steel Authority of India, TATA STEEL, amcor, ALCOA, ArcelorMittal, RioTinto, SOLVAY, BAOSTEEL
Chemical	ABB, PPG, Ciba, TATA CHEMICALS LIMITED, BAYER, AkzoNobel, BASF
CPG	Unilever, TESCO, STARBUCKS COFFEE, KRAFT, Timberland, PEPSICO, PUMA, WRIGLEY, Arla
Standards Bodies	ISO, SUSTAINABILITY CONSORTIUM, UNEP, THE Keystone CENTER, CARBON DISCLOSURE PROJECT ACCREDITED PROVIDER, The Greenhouse Gas Protocol Initiative

GaBi I-report

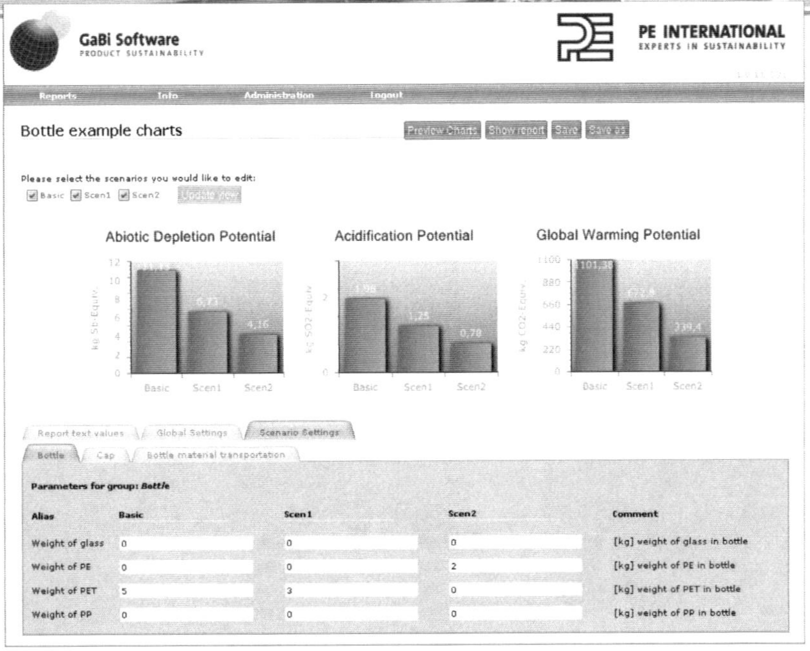

Vision for Sustainable Packaging

Five Winds INTERNATIONAL

PE INTERNATIONAL

Definition of Sustainable Packaging

SUSTAINABLE PACKAGING COALITION®

In 2005, the SPC completed version 1.0 of the Definition of Sustainable Packaging. This definition represents an important first step in articulating a common understanding of the term "sustainable packaging." It provides a common vision and a framework for understanding activities directed toward improving packaging, and continues to inform the future vision of the coalition and its individual member-companies. This definition has been widely adopted throughout the packaging industry.

Sustainable Packaging:

- Is beneficial, safe & healthy for individuals and communities throughout its life cycle;
- Meets market criteria for both performance and cost;
- Is sourced, manufactured, transported, and recycled using renewable energy;
- Optimizes the use of renewable or recycled source materials;
- Is manufactured using clean production technologies and best practices;
- Is made from materials healthy in all probable end of life scenarios;
- Is physically designed to optimize materials and energy;
- Is effectively recovered and utilized in biological and/or industrial closed loop cycles.

Together, these criteria characterize our vision of sustainable packaging. No ranking is implied in the order of definition criteria.

Five Winds International
PE INTERNATIONAL

Determine your Vision

Use the right tool

Think long-term

Where are you on the Leadership Maturity Curve?

Five Winds International
PE INTERNATIONAL

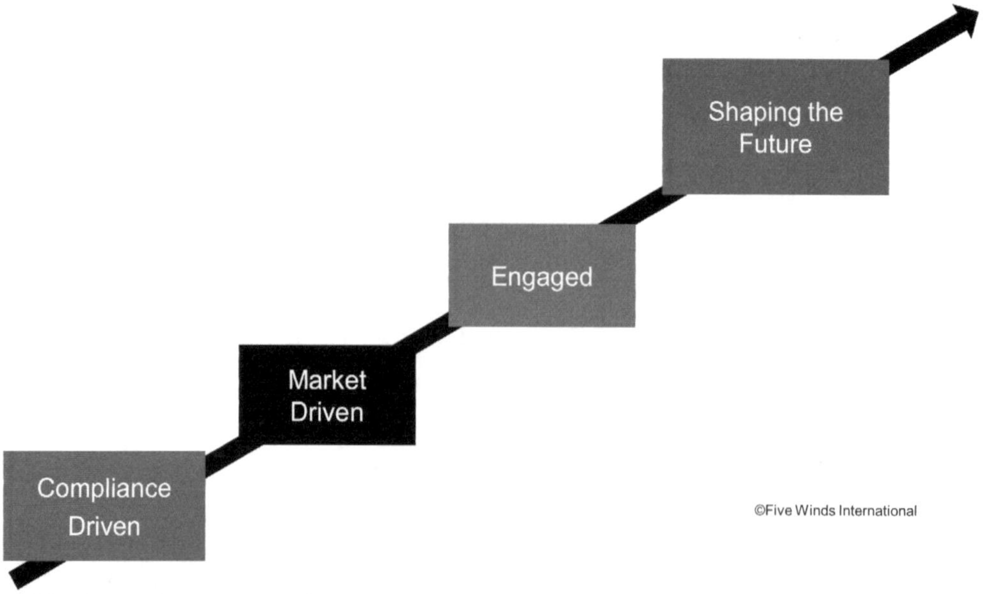

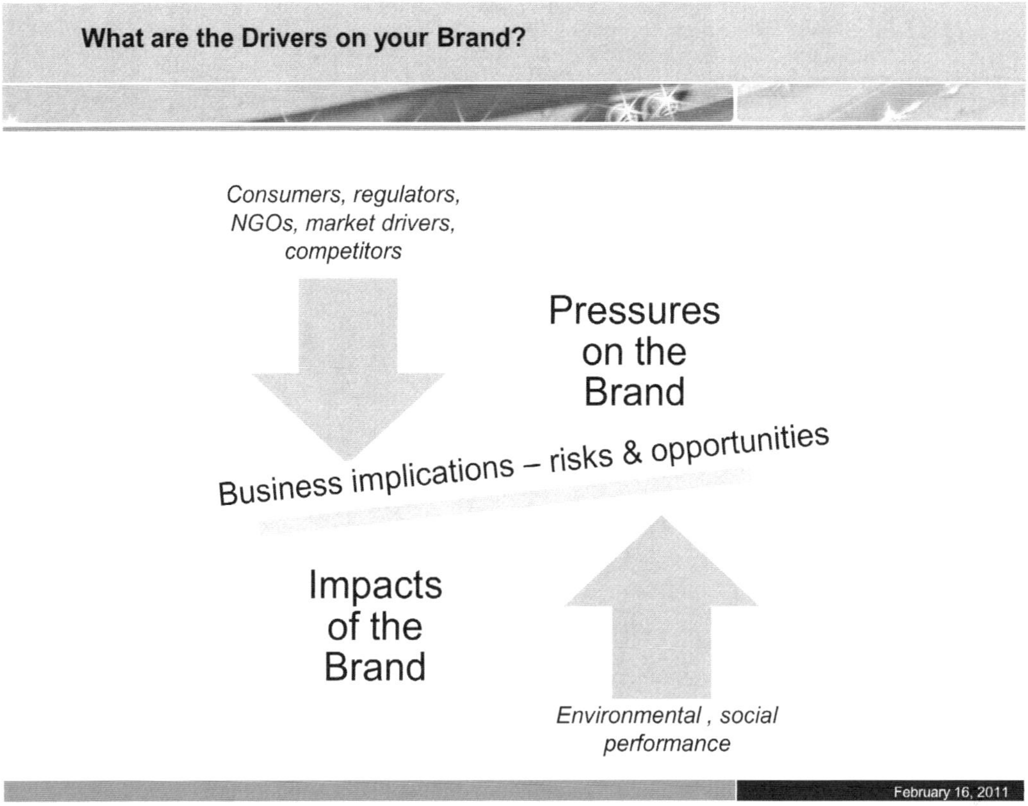
What are the Drivers on your Brand?
Consumers, regulators, NGOs, market drivers, competitors
Pressures on the Brand
Business implications – risks & opportunities
Impacts of the Brand
Environmental , social performance
February 16, 2011

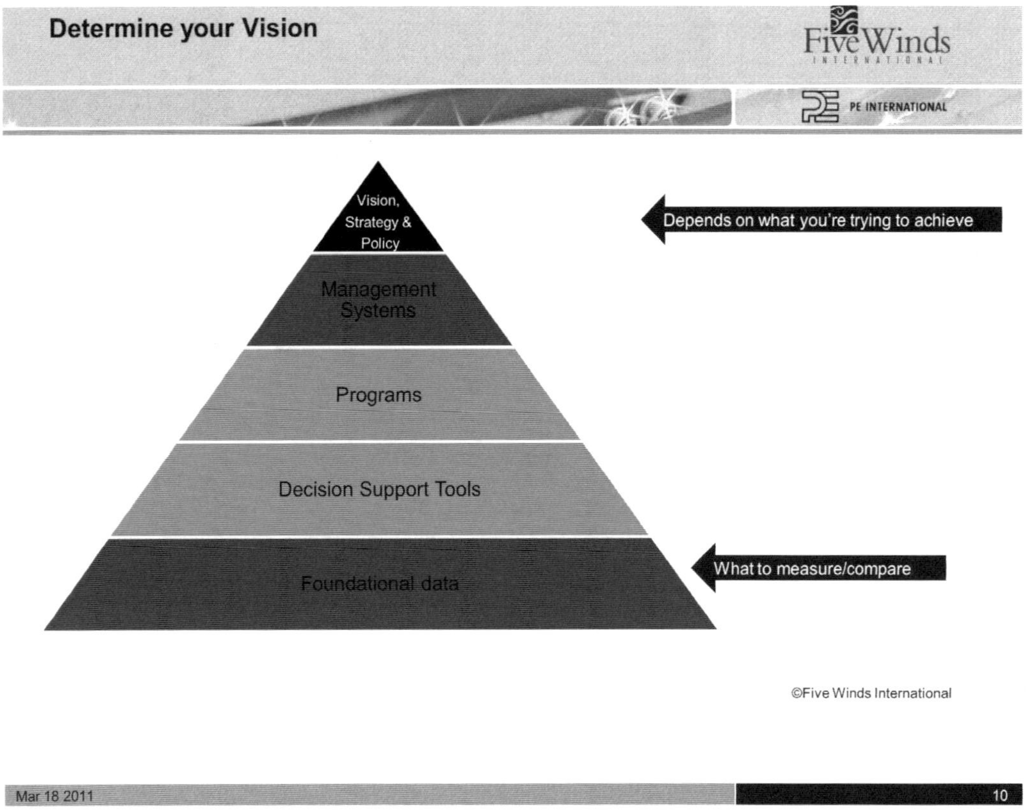
Determine your Vision
FiveWinds INTERNATIONAL
PE INTERNATIONAL
Vision, Strategy & Policy
Management Systems
Programs
Decision Support Tools
Foundational data
Depends on what you're trying to achieve
What to measure/compare
©Five Winds International
Mar 18 2011
10

Use the Right Tool

Five Winds International | PE International

Increasing level of detail & data

- **Checklists**
- **Product scorecards**
- **Directional tools**

- **LC Screening**
- **Hot spot analysis**

- **LCA**

Life Cycle Screening

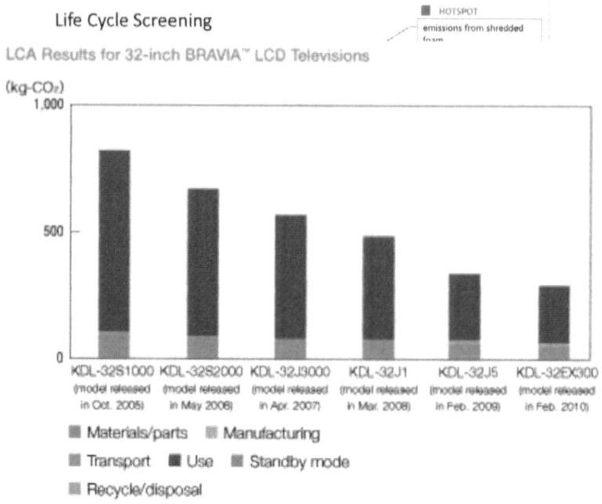

Application of Tools

Five Winds International | PE International

- Use LCA to:
  - Understand the impacts of packaging from raw materials to end of life
  - Understand where packaging impacts fall in comparison to product impacts

  → *drive meaningful improvements in multiple performance indicators across the life cycle*

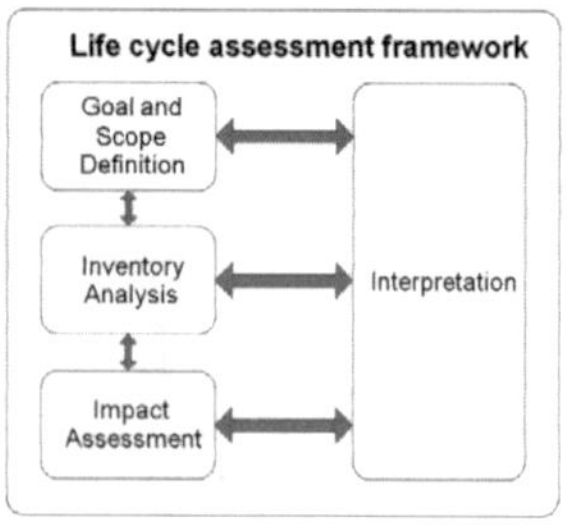

- Use innovation to:
  - Clarify the purpose of the packaging and optimize to it
  - Find ways to leave a positive impact as a result of your packaging – reduce the impacts of other life cycle stages

  → *Reduce more waste/ energy than you require to make the packaging*

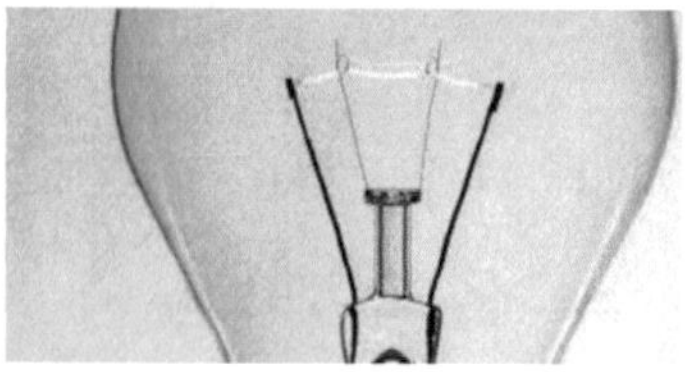

Think long-term

Five Winds International

PE INTERNATIONAL

?

Mar 18 2011 13

Succeed in the Marketplace

RECYCLABLE

compostable

FSC

USDA CERTIFIED BIOBASED PRODUCT

PRODUCT 57%

SCS CERTIFIED

Minimum 74% Recycled Content

37% Post-consumer

37% Pre-consumer

SCIENTIFIC CERTIFICATION SYSTEMS

Environmental Facts

Overall Weighted Score .... 6 / 10

Energy

Embodied energy 2,800kWhr

Type of energy used: 2,000kWhr coal, 800kWhr solar PV

Energy usage, avg. est. 1,900kWhr/yr

Transportation origin

Product: USA

Materials: USA, China, Korea, South Africa

Resources

Product

Mass 10kg

Non-virgin material 5%

Recyclable/Compostable material 30%

aluminum, steel, plastic #1

Ingredients: Polyethylene terephthalate (PET), aluminum, steel, glass, copper, fiberglass, acrylonitrile-butadiene styrene (ABS), lead-free solder, nematic liquid crystals, polyimide, indium-tin oxide, Polycarbonate, Poly(methyl methacrylate) (PMMA), Styrene-butadiene co-polymer, Polyethylene ether, Triphenyl phosphate, polybrominated flame retardant, silicon, silicon dioxide, silicon nitride, selenium, cadmium, antimony, dopants

Life Expectancy 4-7yrs

End-of-life .... return to manufacturer

Packaging & Misc.

Mass 800g

Non-virgin material 20%

Recyclable/Compostable material 100%

cardboard, paper, PLA plastic

Ingredients: cardboard, paper, PLA plastic, soy-based ink

End-of-life .... recycle, compost

Toxins

Restricted/Toxic ingredients: polybrominated flame retardant, cadmium, antimony, dopants

Restricted/Toxic production waste: toluene, mercury oxides, cadmium, antimony, arsine, silane, chlorine, phosgene, perfluorocompounds (CF_4, C_2F_6, NF_3, SF_6, CHF_3)

Water

Embodied water 2,600L

Water pollution 2/10

Social

Labor Practices 8/10

Fair trade 4/10

Transparency 6/10

Mar 18 2011 14

Summary

Five Winds INTERNATIONAL

PE INTERNATIONAL

Determine your Vision

Use the right tool

Think long-term

Mar 18 2011 15

PE INTERNATIONAL
EXPERTS IN SUSTAINABILITY

Laura Flanigan
l.flanigan@fivewinds.com
773 698 8073 ext 1

GaBi Software Five Winds INTERNATIONAL

SoFi Software PE INTERNATIONAL EXPERTS IN SUSTAINABILITY

Registered trademarks of PE INTERNATIONAL AG, Germany

PE Five Winds Inc.

PE Five Winds is the North American presence of PE INTERNATIONAL. In January 2011, Five Winds and PE, two sustainability leaders in product and corporate sustainability software and strategic consulting services, joined forces. With more than 20 years of experience, PE Five Winds enables businesses to understand their impact on the environment, improve company performance and succeed in the market place.

SunChemical®
a member of the DIC group

Unintended Consequences – Ink Technology & Environmental Footprint

Working for You

Sustainability Packaging Symposium
18th March 2011, Chicago, USA

Market Leadership

1818 – Sun Chemical's roots begin with Lorilleux, one of the first ink producers

1929 – Four ink companies merge to form General Printing Inks (GPI)

1936 – First pigmented flexo inks

1945 – GPI changes its name and Sun Chemical is born

SunChemical

1957 – Ansbacher-Siegle merges with Sun Chemical to form today's Performance Pigments division

1958 – Development of solvent-based flexo inks on polythelene films

1969 – Sun Chemical patents first UV inks

1987 – DIC acquires Sun Chemical

1999 – Sun Chemical offers digital inkjet ink

2010 – Sun Chemical opens state-of-the-art food packaging plant in Frankfurt and sheetfed ink hub in Dahej, India

Strong Parent Company

$8.1 billion net sales
204 subsidiaries and affiliates
22,583 employees in **64** countries

Color & Comfort by Chemistry

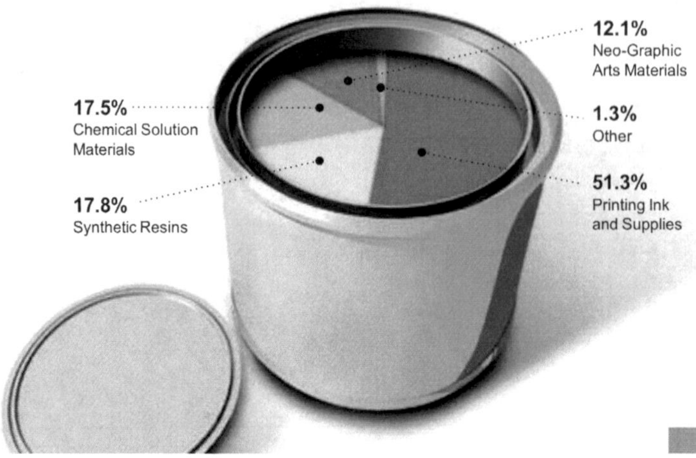

Our Sustainability Policy

- Eco-efficiency
  - Is achieved by delivery of **economically** competitive goods and services that satisfy our customers' needs and brings quality of life
  - While progressively reducing ecological impact and resource intensity throughout the life-cycle
- Road Map – Starting a journey in sustainability
  - A structured and robust approach for processes and products
  - Entitlement to set targets for improvements
- Metrics
  - Key data measured and published regularly

4

Sun Chemical Sustainability Reports

5

Unintended Consequences of a Sustainability Policy

Customers may change technology based on a sustainability perception and not facts

Sustainable Technology

Case Study 1 :

A label printer using conventional offset technology is losing business against water based flexo printers…

On the grounds of more environmental friendliness of the water based inks…

Print Technology Selection: Packaging

Substrate

- Type: flexible film, paper, carton board
- Form: roll to roll, roll to sheet, sheet to sheet
- Lamination or surface printing

Deliverables

- Print quality (mechanical and color)
- Food safety (migration resistance)
- Water or chemical resistance
- Special Effects (tactile, pearlescent, metallic…)

Printing Process

- Run length
- Press performance (make ready, washing frequency, cleaning…)
- Drying process: oxidation/hot air drying/energy cure

Customers consider many factors when selecting a printing technology

8

Ink raw material choices depend on print & application needs

- **Pigments**
  - Provide color - mainly petrochemical-derived - typically 10 - 20% of the ink content
  - Naturally-derived pigments do not afford the required color values or lifetime
- **Resins**
  - Provide gloss, adhesion, viscosity, rub resistance, durability, pigment dispersion
  - Bio- vs. petroleum-derived resins must be optimized to achieve desired properties
  - Migration properties must be designed for challenging applications like food packaging
- **Solvent**
  - Provides mobility for solid components and substrate penetration
  - Offset inks can contain mineral oils, petroleum distillates, vegetable oils or a mixture
  - Key issues are: (1) safety in use (2) energy management & recovery (3) whether from a renewable resource (4) free of volatile organic compounds (VOCs) and (5) residual content
- **Additives**
  - For special performance needs (e.g. adhesion)
  - Migration properties and interactions must be considered

Sun Chemical Key Sustainability Criteria for Products

- **Sustainability is about developing and manufacturing products which minimize the total environmental impact; not just the environmental impact of individual supply chain participants**
- **Sustainable products are those that have one or a combination of environmental benefits:**

> - Conserve energy and material resources in the printing process
> - Decrease carbon footprint in ink raw materials, ink production, substrate production, or the printing process
> - Use renewable / bio-based materials
> - Are recyclable or compostable
> - Reduce VOC emissions
> - Reduce packaging or lightweights packaging
> - Reduce substrate waste

Sun Chemical Products & Bio-based Raw Materials

Ink Type	Bio-based Materials* (%)	Sun Chemical Position
Coldset Inks	5 – 40%	• Sun Chemical leads with 39% bio-based content versus standard 7%. • First to bring soy-based inks to market.
Heatset Inks	25 – 35%	• Standard is 28% bio-based content.
Sheetfed Inks	40 – 77%	• SunLit™ Diamond leads with Bio-Renewable Content (BRC) Index of 80 certified by NAPIM.**
Publication Gravure Inks	30 – 40% and 65% for water-based	• Sun Chemical has developed a water-based version.
Energy Curing Inks	Typically less than 5%, up to 20%	• SunCure® Hy-bryte® leads with 20% bio-based content.
Packaging Inks - Solvent-based	20%	• Main resin – nitrocellulose – is naturally derived. Other bio-based content available. • Inks suited for bio-substrates.
Packaging Inks - Water-based	Typically 0%, up to 20%	• Bio-based content available, such as starches.

*Estimated content from material source. Sheetfed inks measured with ASTM 6866 Carbon 14 Protocol.
** National Association of Printing Ink Manufacturers.

Life Cycle Thinking: Intensive agriculture can lead to environmental impact similar to that of petroleum derived products (machinery, fertilizers, biocides), not to mention rainforest destruction and shift of land from food production to industrial use.

Sun View: Bio-derived raw materials use should be maximized not because they are from vegetable source but because they are from a renewable resource that can become sustainable if managed properly

11

Sun Chemical Products & Gate-to customer gate Carbon Footprint

Summary of Product Carbon Footprints (gCO_2e per Kilogram Product Delivered)

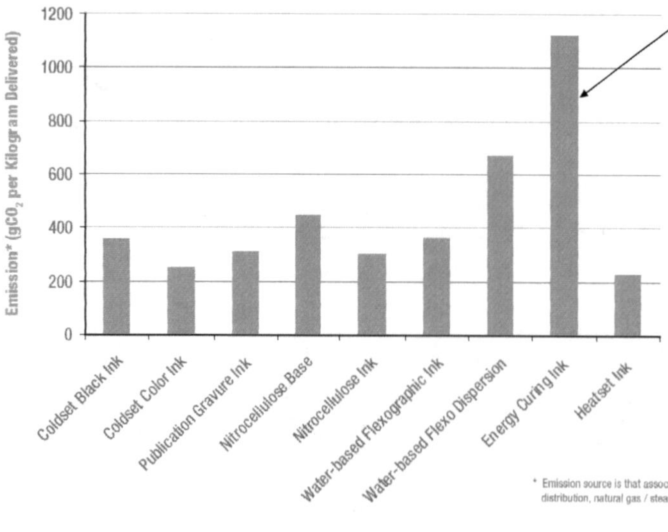

Life Cycle Thinking: Other benefits must outweigh the carbon footprint.

Sun View: Gate-to-Gate Carbon Footprint for Sun Chemical products are to be used for:

•Input into larger Life Cycle Analysis

•For Sun to decide target actions for Carbon Footprint reduction in Sun production/distribution

Results should not be used directly for one to one technology comparison

* Emission source is that associated with average distribution, natural gas / steam, electricity and other fuel.

12

Carbon Foot printing Inks is complex

Challenges for ink makers Life Cycle Analysis:

- Thousands of raw materials involved upstream
- Various & complex print technology downstream

Water-based liquid inks contain 50-60% of water hence a lower relative impact of raw materials to be expected

On-going actions:

- Connection of Competence with PrintCity and other Industry Groups
- Sun's suppliers survey
- Product development focusing on waste and energy reduction at customers

Gate-to-Customer Gate carbon dioxide equivalent emissions from printing ink based on analyses and review of GHG LCA information sources:

- 280 gCO_2e per kilogram (130 gCO_2e per pound) of product delivered (manufacturing)
- 2 $kgCO_2e$ per kilogram (0.91 $kgCO_2e$ per pound) of ink delivered including supply chain

De-inking Industry view

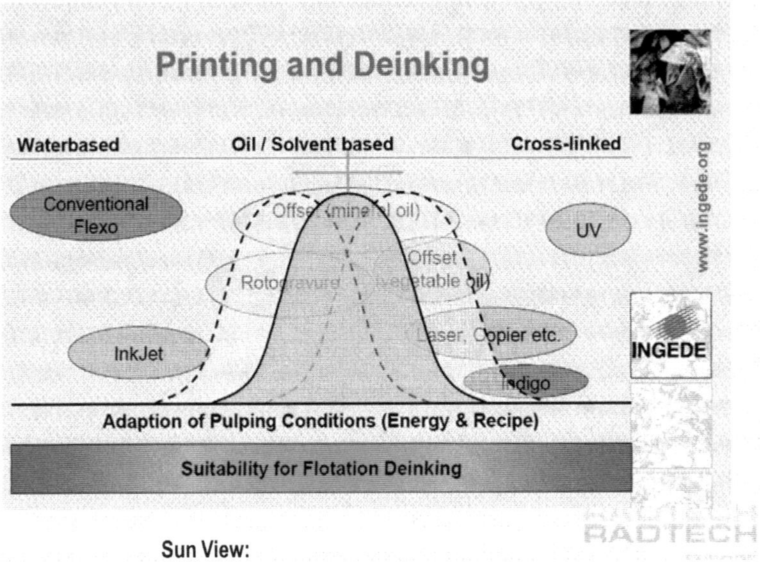

Life Cycle Thinking:

- Vegetable inks more difficult to de-ink than mineral oil inks
- Water based inks can be made easier to de-ink but at the expense of press performance.
- Many types of inks have a calorific value greater than coal or wood

Sun View:

- Recycling ≠ Downcycling
- Sewerage ≠ Landfilling ≠ Composting

14

Drying Process (1/2)

Offset inks

Life Cycle Thinking:

- Less energy is required to oxidatively dry an offset ink than a water-based flexo ink
- Oxidative drying.requires zero energy to set the ink

Water Based Flexo Ink

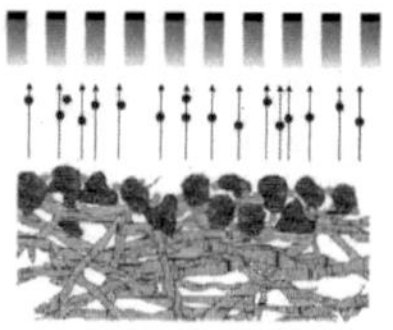

Life Cycle Thinking:

- Water has a high latent heat of vaporization and requires more energy to evaporate, particularly on non-absorbent substrates than solvent or oxidatively dried inks
- Can be compensated by printing lower film weights

Sun View:

- Continually develop quicker and more efficient drying ink systems to improve press performance (speed), print quality, and process efficiency (e.g. improved work and turn).

Case study 1 : Sustainable Technology - Conclusion

Case 1

- A label printer using oxidative drying offset technology is losing business against water based flexo printers...
- On the grounds of increased environmental friendliness of the water based inks...

Facts

- Oxidatively dried offset inks use bio-derived raw materials, require zero drying energy and are easier to de-ink
- Water Based Inks exhibit a lower raw material carbon footprint

Sun Chemical view

- The initial technology choice is still mainly led by cost efficiency and print quality
- No single print and ink technology stands out as better than the others regarding the total environmental footprint. The environmental footprint is highly complex
- Life Cycle Analysis data may also not be available for fair comparison and methodology and scope of analysis can vary widely.

Unintended Consequence: Loss of sustainability based on abandonment of bio-derived raw materials, increased drying energy, and adverse environmental impact due to less efficient de-inking. Unsubstantiated claims for environmental improvement.

16

Safer Food Packaging

Case Study 2 :

- BBC News, 8th March 2011: « Food sold in recycled cardboard packaging 'poses risk' »
- The Telegraph, 8th March 2011: « Cereal box health warning after recycled cardboard study. Breakfast cereal manufacturers are to stop using recycled cardboard in packaging after a study indicated that current boxes could pose a cancer risk. »

The "Mineral Oil" case

Background: Studies (Grob, Zürich) found "mineral oil" from offset printing inks in food:

- Recycled board in Europe contain insufficiently de-inked pulp containing residual ink mineral oil → **Waste Stream Issue**
- When conventional inks are used for primary packaging and no proper functional barrier is in place some ink constituents can migrate to the food → **Migration Issue**

Reactions:

- UK - media reporting carcinogenic effects of mineral oils
- UK - Food and Drink Federation said the Swiss study was "a good starting point for further investigations", but insufficient in itself to justify discontinuing the use of recycled card
- Germany - Government told the food and packaging industry to take immediate steps to reduce the risk from mineral oil and is considering mandatory rules
- UK - Food Standards Agency is performing independent research & some leading food manufacturers are voluntarily changing their packaging
- Cereal firm Jordans stopped using recycled board. Kellogg's and Weetabix are taking actions to reduce mineral oil content in their packaging.

18

HOW DOES MIGRATION OCCUR?

1.	Physical Migration	**PENETRATION MIGRATION** Migration from the printed side through the substrate onto the unprinted side.	 Penetration through the substrate
2.	Physical Migration	**CONTACT MIGRATION** Migration from the printed side to the unprinted side of another sheet in a stack or roll.	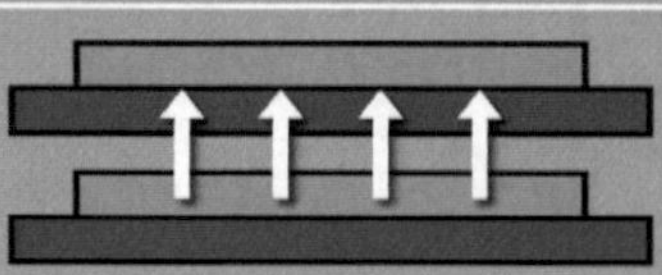 'Set-off' transfer onto the reverse side in the stack
3.	Gas Phase Migration	**EVAPORATION MIGRATION** Migration due to the evaporation of volatile materials by heating (e.g. cooking, baking, or boiling frozen products in their original packaging).	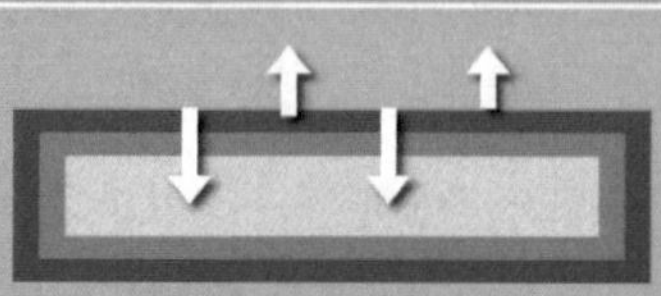 Vapour phase transfer
4.	Gas Phase Migration	**CONDENSATION MIGRATION** Migration through steam distillation during cooking, baking or sterilisation.	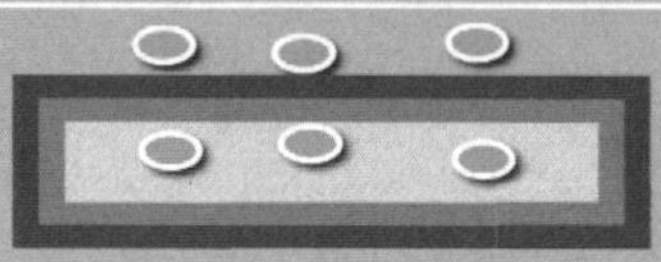 Condensation extraction

What are the sources?

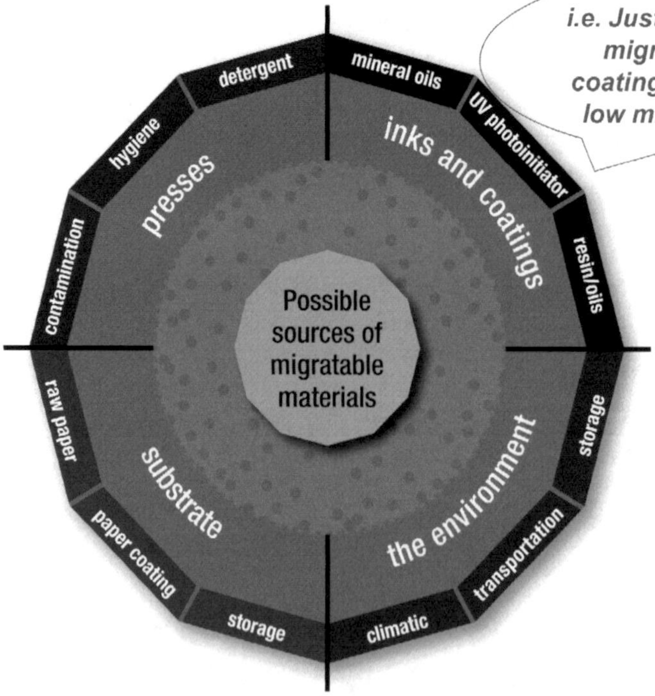

i.e. Just switching to low migration inks and coatings will not ensure low migration printing

20

The complete package needs to be considered....

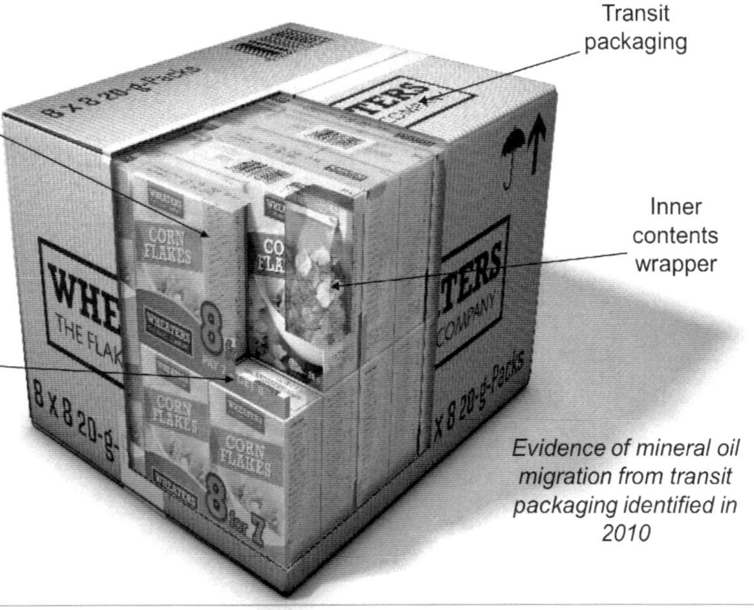

Evidence of mineral oil migration from transit packaging identified in 2010

The unintended consequence of a sustainability-led recycling practice was contamination of food with mineral oil

Unintended Consequences - Ink Technology & Environmental Footprint

- **Case 1:** Unsubstantiated selection of water based ink over offset ink
- **Case 2:** Environmentally driven recycling program resulting in food contamination with mineral oil

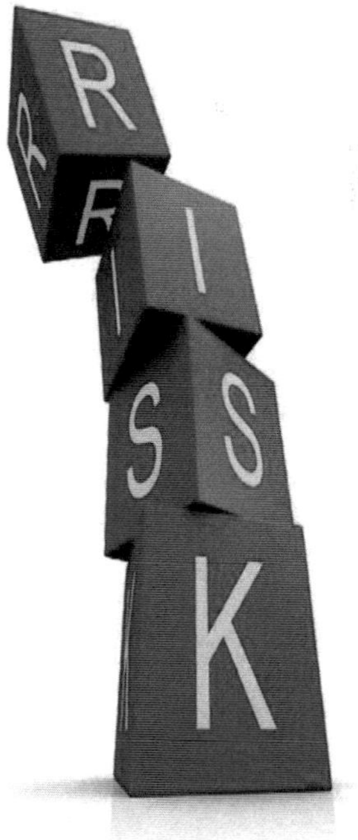

Other examples:

- The migration from the use of petroleum derived solvents to bio-ethanol has had the unintended consequence of diminishing the food supply, especially to third world countries
- A cereal manufacturer replaced a high cost, high carbon footprint foil liner bag with a lower weight PE liner. The PE absorbed and released many components into the cereal, including 4—methylbenzophenone

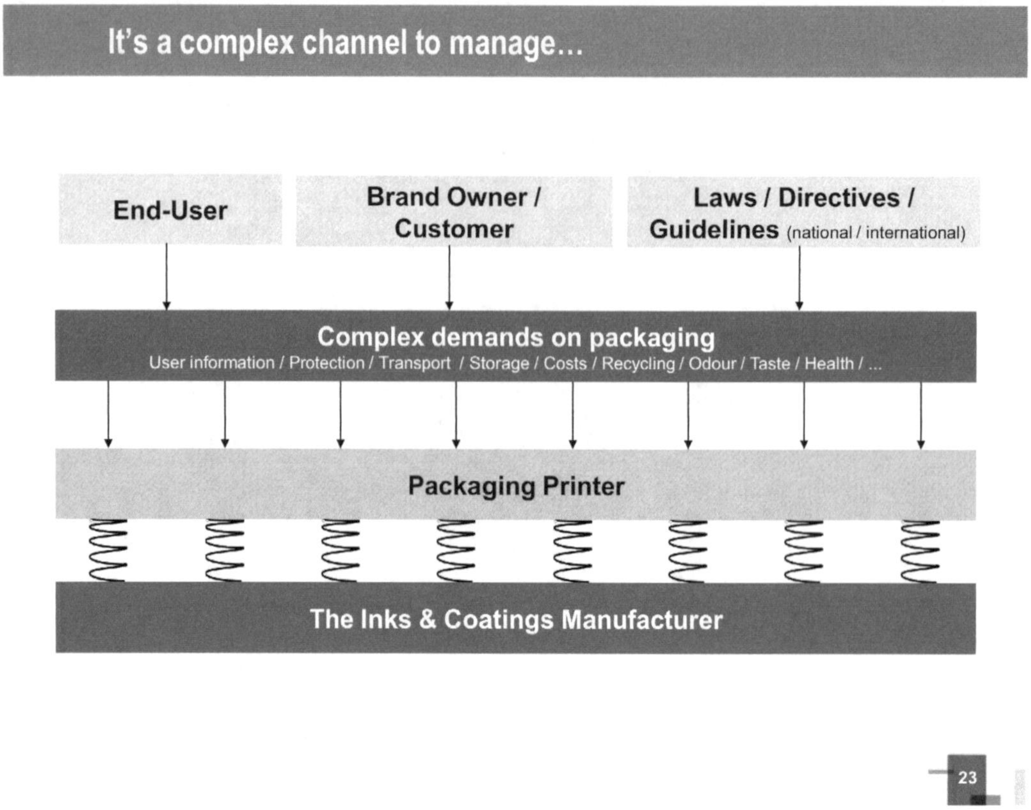
It's a complex channel to manage…
End-User
Brand Owner / Customer
Laws / Directives / Guidelines (national / international)
Complex demands on packaging
User information / Protection / Transport / Storage / Costs / Recycling / Odour / Taste / Health / …
Packaging Printer
The Inks & Coatings Manufacturer
23

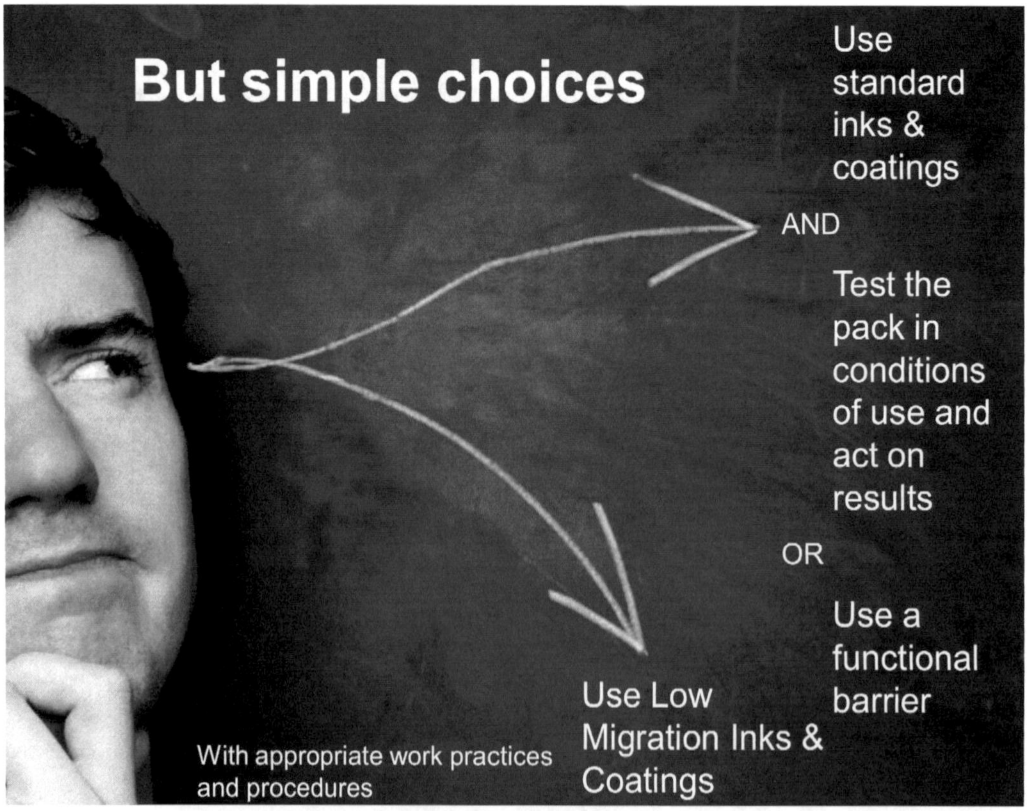
But simple choices
Use standard inks & coatings
AND
Test the pack in conditions of use and act on results
OR
Use a functional barrier
Use Low Migration Inks & Coatings
With appropriate work practices and procedures

Support from Sun Chemical for the supply chain

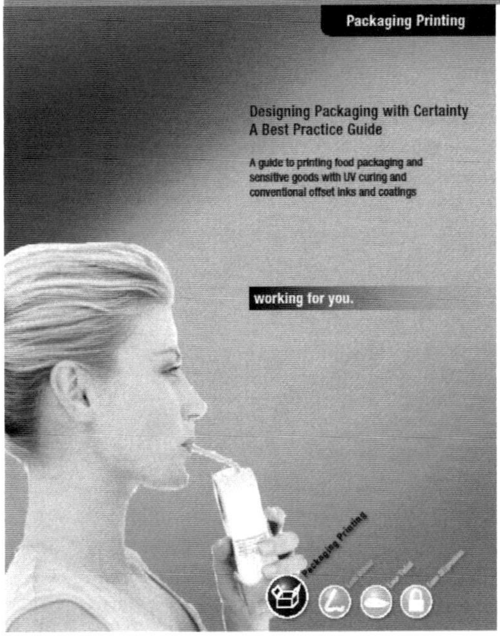

SunChemical

Low migration printing is not just about ink!

4 Step Model to designing packaging with certainty

- Pre-Press & Pack Design
- Selection of materials
- Printing Equipment & Management
- Press Room, Handling, Transport & Storage

25

The best solution is for the packaging supply chain to work together to ensure best possible environmental footprint whilst preserving product performance

26

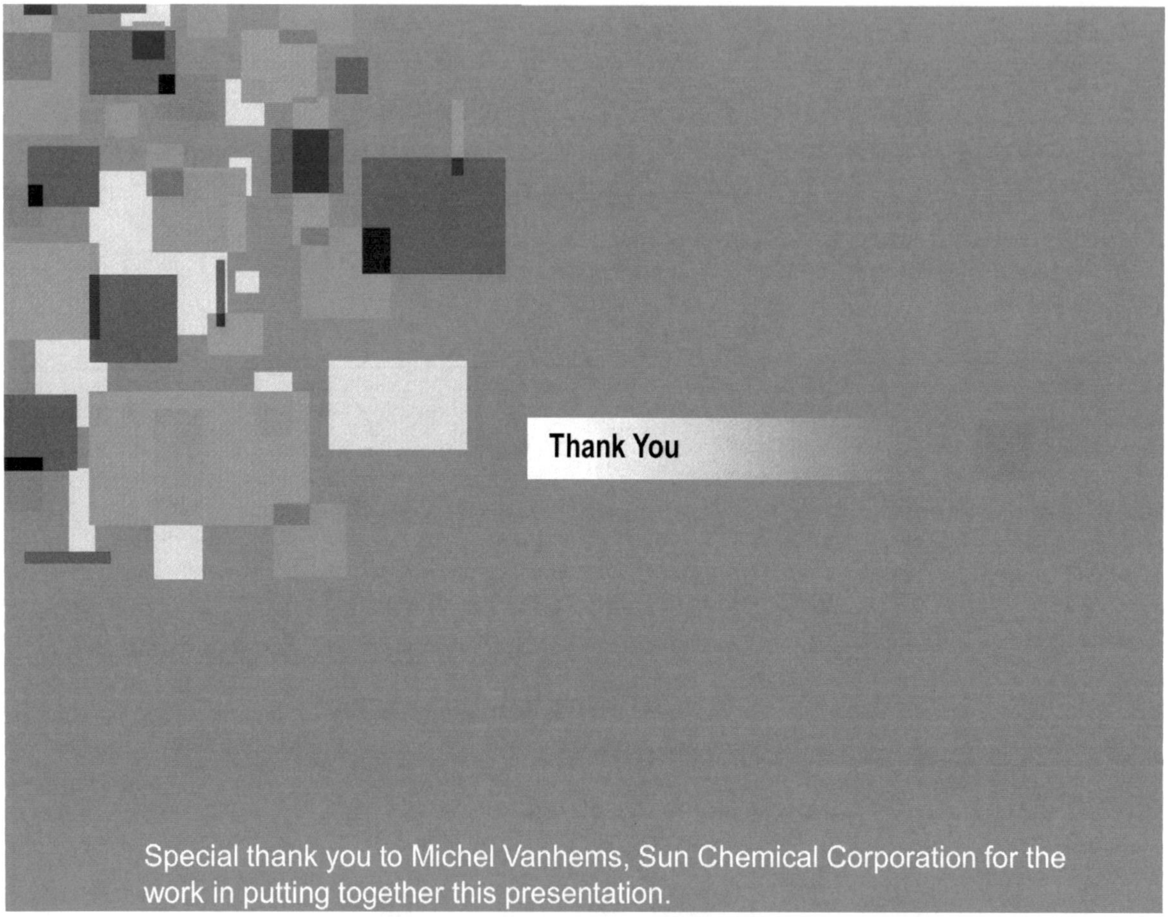
Thank You
Special thank you to Michel Vanhems, Sun Chemical Corporation for the work in putting together this presentation.

It Takes A Village - Using Systems Thinking to Solve Starbucks Recyclable Cup Initiative

Hanna, J. M., Starbucks Coffee Company

Originally presented on: 3/18/2011 14:10:00 - 14:35:00

Starbucks Coffee Company sells approximately four billion single-serve (disposable) cups globally each year. Across the retail industry, approximately 500 billion single-serve cups are sold annually. In the vast majority of communities, these cups are destined for landfill at end-of-life, generally due to lack of market demand for the product (designated as post-consumer poly-coated paper stock and post-consumer thermoformed polypropylene) within the recycling industry.

In 2009, Starbucks set an ambitious goal that all of our single-serve cups would be recyclable or reusable by 2015. Unlike a number of recent marketing campaigns, we do not define "recyclable" based on any particular material that comprises the components of our cups, but rather on our customers' actual access to recycling services in the places where they choose to dispose of the cups. This includes within Starbucks stores, in residential locations, offices and public spaces. This definition of recyclable is consistent with the Federal Trade Commission's Green Guides, which indicate that a product should not be called recyclable unless a *significant majority* of users have readily-available access to recycling service for the product.

Starbucks engaged the assistance of Peter Senge and his team of systems experts at MIT to define the boundaries of the lifecycle of our cups and engage the stakeholders from the entire value chain to develop a systems-based set of solutions to test and, ultimately scale, to create the necessary market drivers to achieve our 2015 goal. At the second of two "cup summits" to date, held most recently at MIT in April 2010, stakeholders including raw material suppliers, primary manufacturers, retailers (including a number of Starbucks competitors) municipal regulators, recyclers, secondary manufacturers and NGOs gathered to understand the impact and often unintended consequences of their individual decisions on the entire system, break down silo thinking, and formulate a series of tests to begin creating the necessary market pull to successfully incorporate used cups from all retailers into their business models.

This discussion will provide an overview and deep dive into Starbucks cup recycling initiative, in the context of using our example for a broader discussion of the unintended consequences of failing to incorporate systems thinking into raw material selection, manufacturing design, purchasing decisions, marketing and policy development.

Although spending time focusing on paper cups may seem trivial in the face of the massive environmental challenges confronting the planet, it is the goal of this discussion to also demonstrate the important role of public perception of where the greatest environmental impacts lie in driving (or inhibiting) one's ability to effectively impact these greater challenges.

Cone

Unintended Consequences in Communicating Sustainability:
Balancing Public Expectation with Product Reality

Mike Lawrence
March 18, 2011

Why Should Your Organization Communicate Sustainability Efforts?

Explosion of Green

3

Consumers are Interested

85%

say their expectations are the same or higher for companies to make and sell environmentally responsible products in the economic downturn

2009 Cone Consumer Environmental Survey

92%

want companies to tell them what they're doing to improve their products, services and operations

2010 Cone Shared Responsibility Study

Consumers are Confused

67%

are confused by the messages companies use to talk about their social and environmental commitments

2009 Cone Consumer Environmental Survey

Increase in Activism and Enforcement

FEDERAL TRADE COMMISSION
Protecting America's Consumers

Privacy Policy

Home | News | Competition | Consumer Protection | Economics | General Counsel | Actions

About Public Affairs | Public Events | Speeches | Testimony | Webcasts | Reporter Resources

ForRelease: 06/09/2009

FTC Announces Actions Against Kmart, Tender and Dyna-E Alleging Deceptive 'Biodegradable' Claims

Agency Testifies About Efforts to Ensure Truthfulness of Environmental Marketing Claims

The FTC has charged Kmart Corp., Tender Corp., and Dyna-E International with making false and unsubstantiated claims that their paper products were "biodegradable." Kmart and Tender have agreed to settle the cases against them; the case against Dyna-E will be litigated. The FTC announced the cases today in testimony before the U.S. House Subcommittee on Commerce,

environmental LEADER
Energy & Environmental News for Business

February 18, 2011

Environmental Enforcement: EPA Orders Disposal of Seized E-Waste Bound for Vietnam

Share | Email This Post | Add your comments

RELATED STORIES
Companies Pay $20 Billion for Pollution Control Equipment
Seven PepsiCo UK Plants Cease Sending Waste to Landfills
Military Fuel Use Up 175% Since Vietnam

Two recycling companies have been **ordered to properly dispose of computer waste** they attempted to illegally export from Minnesota to Vietnam, the U.S. Environmental Protection Agency announced Tuesday.

The EPA alleges that the companies attempted to export hundreds of computer monitors to Vietnam for disposal, through the Port of Seattle. The agency is imposing a $31,600 penalty against the companies for violating federal hazardous waste laws.

According to the EPA, Toronto-based Metro Metals Corp., and Avista Recycling, Inc., a recycler based in Minnesota, arranged for the export of a shipment of 913 discarded computer monitors to Vietnam in early December.

AdvertisingAge

Agencies Beware: Culpability Potholes Ahead

FTC's Updated Guidelines, New Court Precedents Indicate Firms Must Step up Due Diligence in Their Work

By: Rupal Parekh Published: March

TriplePundit people, planet, profit

Shareholders File Record Number Of Environment Related Resolutions

By Gina-Marie Cheeseman | February 22nd, 2011

FTC Proposed Changes and Implications for Companies

What are the Green Guides?

- Part of FTC's responsibility to stop "unfair methods of competition," and "unfair or deceptive acts or practices"
- Rules to help marketers ensure the environmental claims they are making are true, substantiated and not misleading
- Applies to all forms of environmental marketing: advertisements, labels, package inserts, promotional materials, logos, product brand names and digital or electronic media, including Internet
- **Guidance covers topics including:**
  - Recyclable
  - Made with renewal materials/renewable energy
  - Carbon offsets
  - Free-of/non-toxic
  - Certifications/seals of approval
  - Ozone-safe
  - Degradable
  - Compostable

Implications for Companies

Companies are responsible for whatever consumers believe their marketing messages mean

- FTC warns marketers "not to make unqualified general environmental benefit claims" because "very few products, if any, have all of the attributes consumers perceive" from such claims.
- FTC "requires marketers to substantiate every...implied environmental benefit that consumers reasonably could take from such a claim."
- Marketers "...should ensure that the advertising's context creates no deceptive implications."

Implications for Companies

Companies must provide supporting information at point-of-sale

79%
want companies to provide information about environmental commitments on the package to help them make informed purchases

2011 Cone Green Gap Trend Tracker

ftc
eco in the market

"Any disclosures needed to prevent an advertisement from being misleading must be clear and prominent and in close proximity to the claim...Websites cannot be used to qualify...because consumers likely would not see that information before their purchase."

Implications for Companies

Companies must provide supporting information at point-of-sale

Smartphone Label Readers

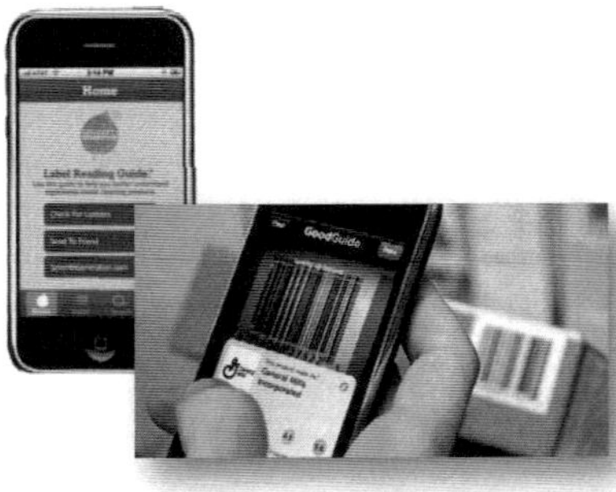

On-pack Information

"This package is suitable for industrial composting. However, industrial composting isn't available in all areas, so check to see if such a facility exists in your community."

Cone | Confidential and Proprietary | 11

Unintended Consequences: Case Studies

Case Study: Nissan Leaf

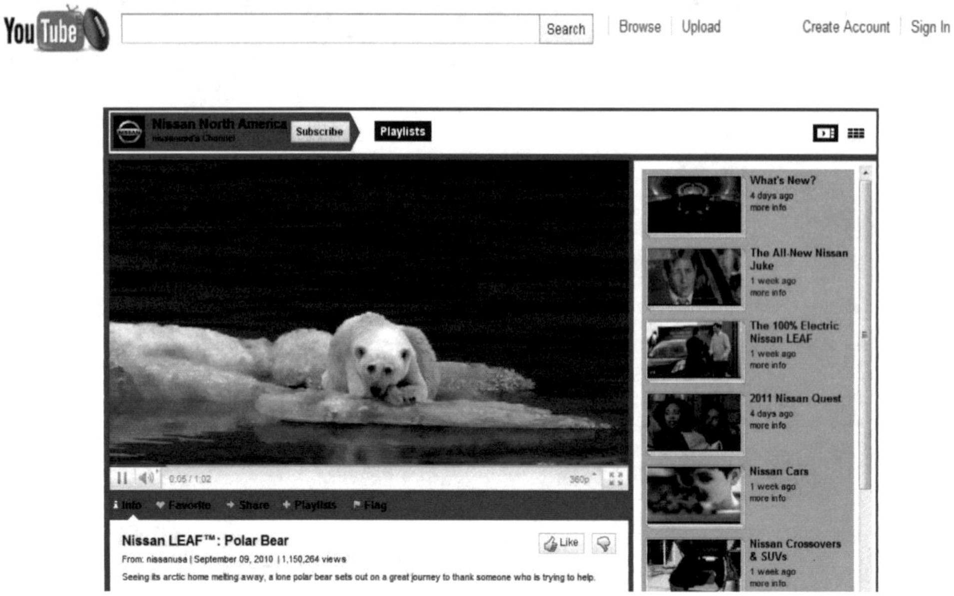

Case Study: SC Johnson

51%

believe a certification on-pack indicates the product has been reviewed by a credible third party

2011 Cone Green Gap Trend Tracker

Milwaukee Journal Sentinel, 2/9/11

"Litigation...contends SC Johnson is deceptively implying that Windex and Shout have been tested by a neutral 3rd party and found to be environmentally friendly."

SC Johnson response

"We're very proud of our accomplishments under the Greenlist system."

Case Study: Ben & Jerry's

Cone | Confidential and Proprietary

Images from: http://www.benjerry.com/

How are Messages Created?

How are Messages Created?

1. Identify Strengths & Points of Differentiation
2. Assess How These are Relevant to Consumers
3. Determine Consumer Benefit of Caring
4. Outline Story to Make Information "Stick"

What is Your Role in Creating Messages?

What is Your Role in Creating Messages?

- **Provide data** relevant to the products and services of your company.
- **Help translate** the impact and importance of the data into customer-friendly ideas and proof points.
- **Provide insight** into supply chain relationships and innovation process to create a compelling story.
- **Help identify** emerging issues and provide context – mapping back to mitigating risk and avoiding "unintended consequences."

What Does Success Look Like?

What Does Success Look Like?

- **Be Precise:** Make specific claims supported by quantitative facts.
- **Be Grounded:** Don't let aspirational messaging or images create a consumer inference you can't prove. Test what message is "received."
- **Be a Resource:** Provide additional information for consumers where they need it at time of purchase decision.
- **Be Realistic:** Put claims in wider context with humility, sense of shared effort and "work in progress" tone.

Mike Lawrence

Executive Vice President
MLawrence@coneinc.com
617-939-8328
Twitter: @Mike_Lawrence

www.coneinc.com | @ConeLLC | www.coneinc.com/whatdoyoustandfor

GlaxoSmithKline

Using LCA to Design More Sustainable Packaging In the Pharmaceutical Industry

Celia Ponder
AIChE Sustainable Packaging Symposium
Chicago, Illinois
18 March 2011

Who are we?

- We are one of the world's leading producers of prescription medicines, vaccines and consumer healthcare products
- 6.3% of global pharmaceutical sales
- Total company sales: £23.2 billion/ $43 billion
- We lead the way in respiratory and anti-viral medications and vaccines
- Over 100,000 GSK people in 117 countries

Each year GMS produces over 4 billion packs

What do we do?

- **Every second...**
  We distribute more than 35 doses of vaccine
- **Every minute...**
  More than 1,100 prescriptions are written for GSK products
- **Every hour...**
  We spend more than £300,000 ($562,000) to find new medicines
- **Every day...**
  More than 200 million people around the world use a GSK brand toothbrush or toothpaste
- **Every year...**
  Our factories produce 9 billion *Tums* tablets, 6 billion *Panadol* tablets and 600 million tubes of toothpaste

Our flagship products

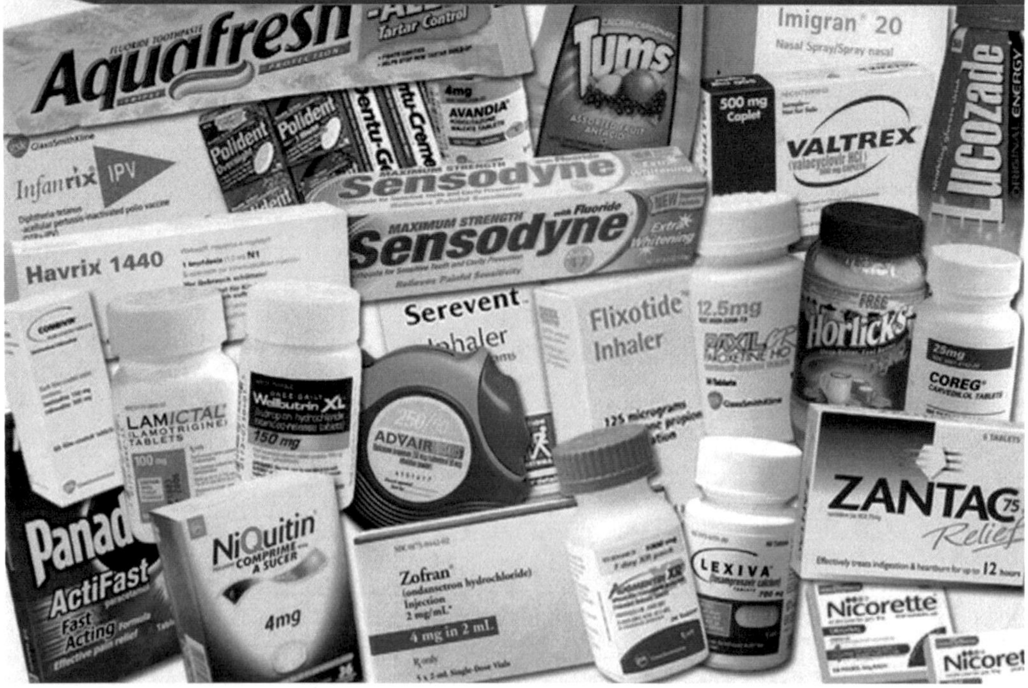

Sustainability by Design

- If we want to make the ***biggest impacts to products, services and costs***, we have to start from the ground up.
- If we want to ***build sustainability into the design of products and services*** we have to think differently about the what and how of R&D.
- Increasing demands and decreasing budgets are likely to mean greater reliance on easily accessible company-wide tools that provide ***early*** assessments and highlight ***sustainability issues***.

Life Cycle Assessment – The very big picture

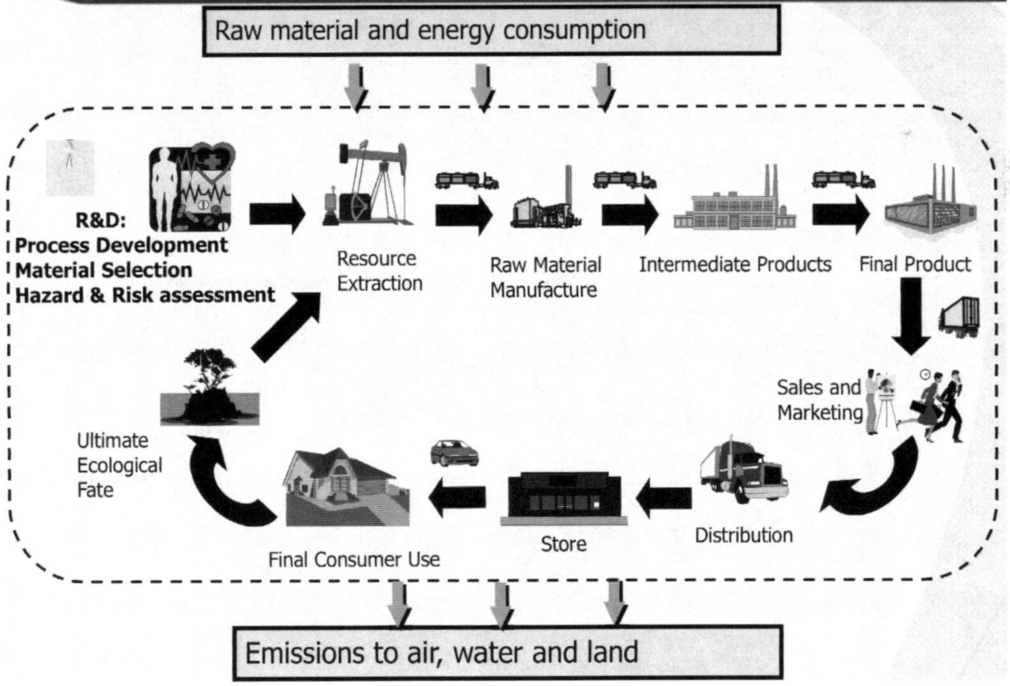

Why is life cycle it important?

- Evaluates the *true greenness of a process*
- Provides a set of *common accepted metrics* to measure specific impacts.
- identifies *trade-offs and hot spots*
- Identifies *opportunities for resource optimisation*
- highlight *hidden costs and impacts*
- provides a *better understanding of the environmental impacts* of processes for a better decision-making for production systems
- Identifies *key impacts* at the life cycle stages of system
- Identifies *information gaps*

GSK Eco-Design Toolkit

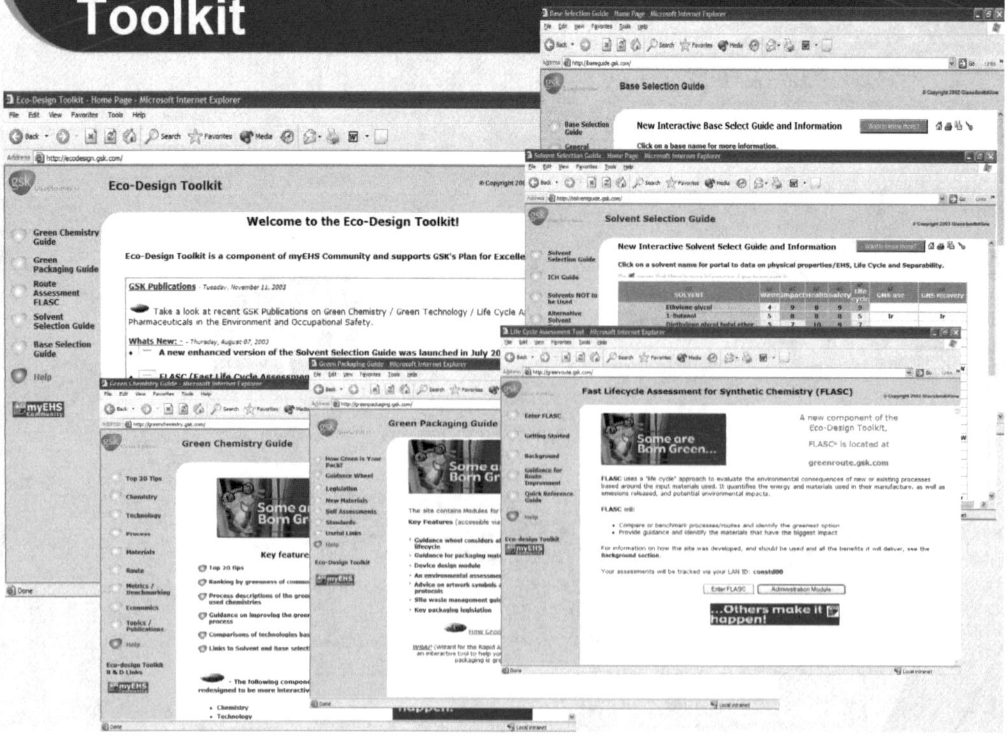

Green Packaging Guide – WRAP (Wizard for Rapid Assessment of Packaging)

- Launched at the beginning of November 2002
  - providing guidance around all aspects of packaging life cycle
- Followed up with a promotional campaign for targeted audiences in Pharmaceuticals and Consumer Healthcare during 2003.
- Launched WRAP Tool in 2003/4
  - Data & Methodology agreed
  - Launched in 2003
- WRAP is an interactive tool that compares a pack against a benchmark covering a range of EHS criteria

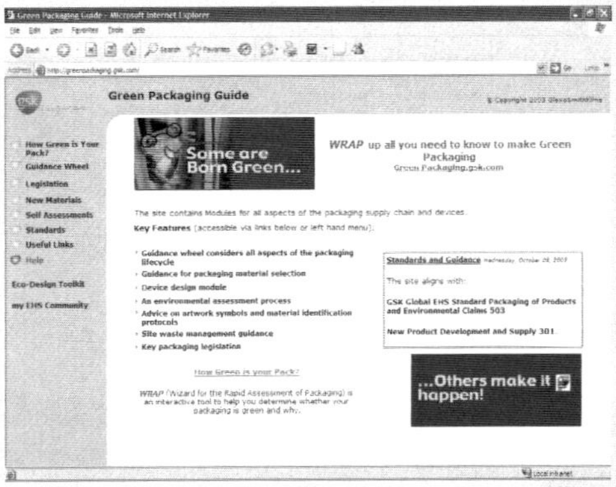

- Green Packaging Guide and WRAP are now being revised to include more materials and life cycle impacts

Example WRAP output comparison

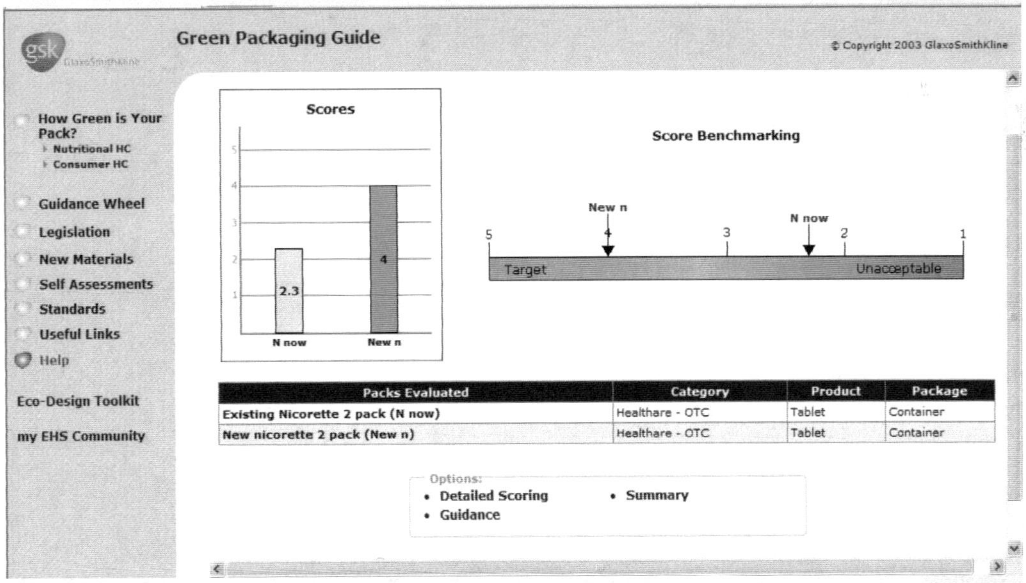

WRAP – Detail behind the scores

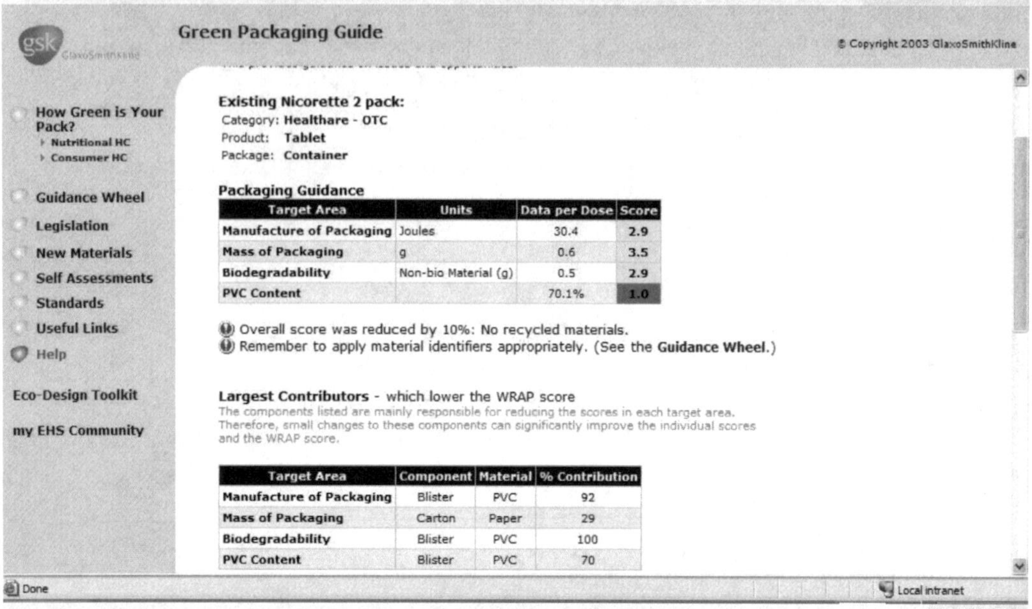
Green Packaging Guide

© Copyright 2003 GlaxoSmithKline

How Green is Your Pack?
- Nutritional HC
- Consumer HC

Guidance Wheel
Legislation
New Materials
Self Assessments
Standards
Useful Links
Help

Eco-Design Toolkit

my EHS Community

Existing Nicorette 2 pack:
Category: **Healthare - OTC**
Product: **Tablet**
Package: **Container**

Packaging Guidance

Target Area	Units	Data per Dose	Score
Manufacture of Packaging	Joules	30.4	2.9
Mass of Packaging	g	0.6	3.5
Biodegradability	Non-bio Material (g)	0.5	2.9
PVC Content		70.1%	1.0

Overall score was reduced by 10%: No recycled materials.
Remember to apply material identifiers appropriately. (See the **Guidance Wheel.**)

Largest Contributors - which lower the WRAP score
The components listed are mainly responsible for reducing the scores in each target area. Therefore, small changes to these components can significantly improve the individual scores and the WRAP score.

Target Area	Component	Material	% Contribution
Manufacture of Packaging	Blister	PVC	92
Mass of Packaging	Carton	Paper	29
Biodegradability	Blister	PVC	100
PVC Content	Blister	PVC	70

Done | Local intranet

Revised Packaging Assessment Tool

- Corporate carbon footprint reduction targets
- Revised to include Carbon Footprint and other life cycle impacts
- Includes materials and optionally transport
- Based on EcoInvent LCI data
- Inputs
  - List of materials
  - Mass of each material
  - Mode of transport
  - Km traveled

Revised Packaging Assessment Tool Output

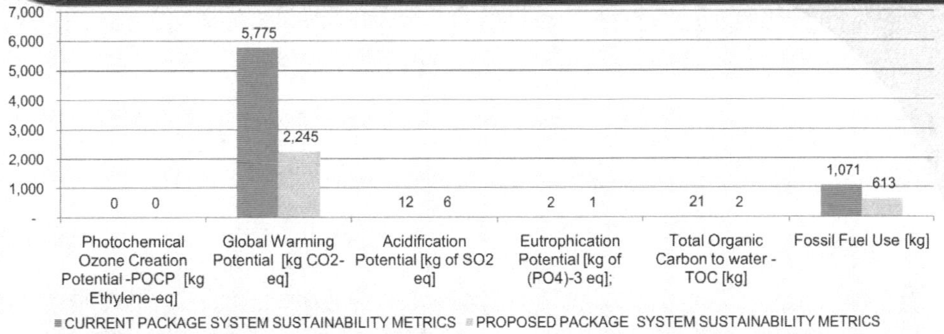

GSK Metrics Packaging and Transport	Total in kg
Raw Material Mass [kg]	2,685
Net energy [MJ]	109,460
Photochemical Ozone Creation Potential -POCP [kg Ethylene-eq]	0
Global Warming Potential [kg CO2-eq]	5,775
Acidification Potential [kg of SO2 eq]	12
Eutrophication Potential [kg of (PO4)-3 eq];	2
Total Organic Carbon to water - TOC [kg]	21
Fossil Fuel Use [kg]	1,071
Water [kg]	80,217

Metric Equivalencies	Total in kg
# homes heated for 1 day	800
km (UK car)	24,014
trees	20
# people's daily water needs	1604

gsk GlaxoSmithKline

Packaging Reduction Case Studies

Reduction of Materials – Carton & Corrugate

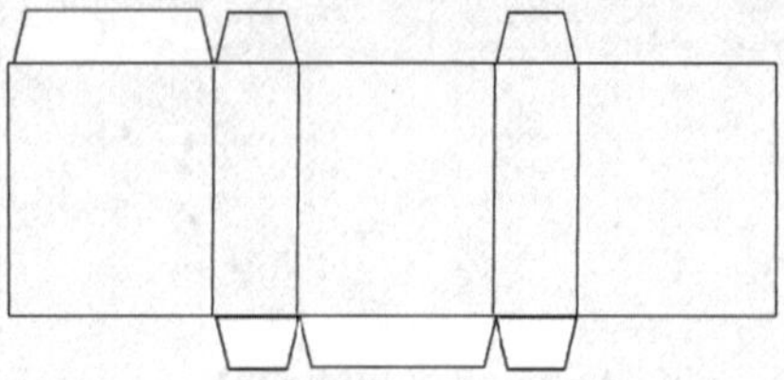

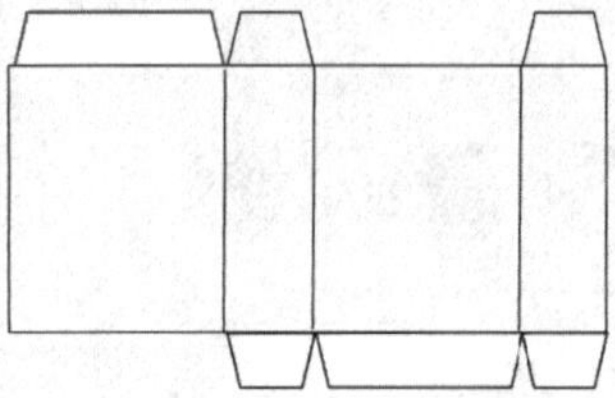

- Carton reduction
  - Old – 5 panel carton
  - New – 4 panel carton
- Larger corrugate box
  - Old – holds 20 cartons
  - New – holds 24 cartons

	Units/ Ship Case	Units/ Pallet	Increased Efficiency
Current Case	20	1600	N/A
Proposed Case	24	1920	20%

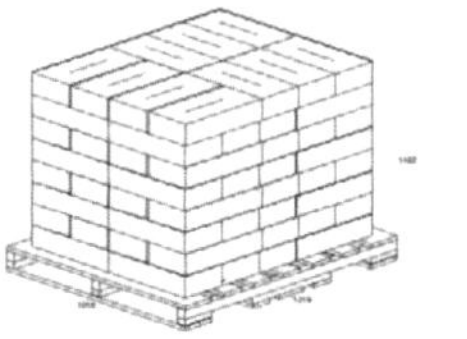

- Current 20/case

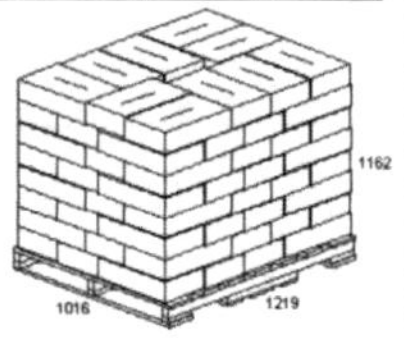

- Proposed 24/case

Reduction of Materials – Carton & Corrugate

GSK metrics	Savings	% Difference
Net energy [MJ]	9,858,265	10%
Total Organic Carbon to water - TOC [kg]	194	9%
Photochemical Ozone Creation Potential - POCP [kg Ethylene-eq]	40	12%
Global Warming Potential [kg CO2-eq]	570,865	10%
Acidification Potential [kg of SO2 eq]	1,325	10%
Eutrophication Potential [kg of (PO4)-3 eq]	144	10%
Oil Equivalents [kg]	95,836	10%

- Overall 10% decrease in life cycle impacts
- The amount of energy saved could heat 600 homes for a month
- The CO_2 emissions saved is the equivalent of taking 22 cars off the road each year, or
- Driving a US car around the world 11 times (www.epa.gov)

Reduction of Materials – Carton & Corrugate

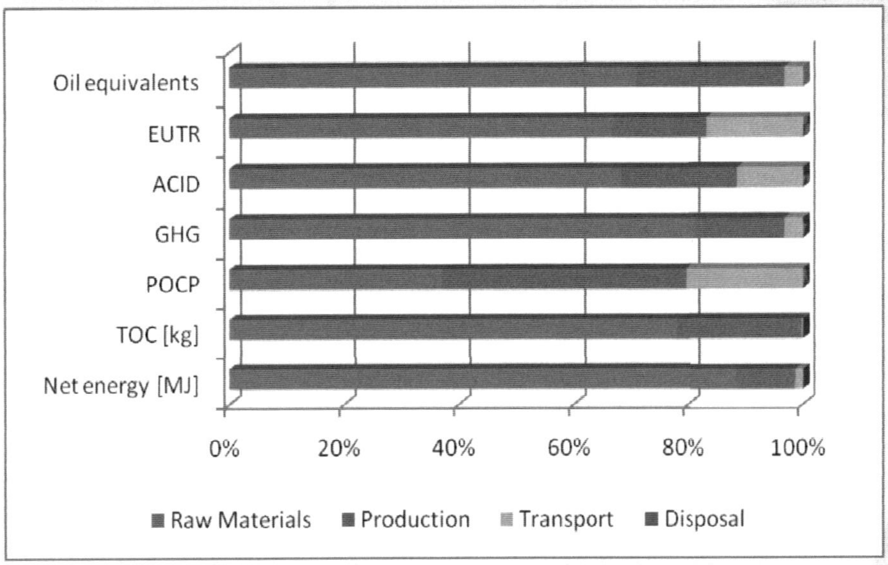

Raw Materials are the largest contributor to life cycle impacts

Reduction of Materials – Horlicks Bag-In-Box

- Laminate change from 3-ply to 2-ply

- Display carton reduction
  - Reduction in dimensions
  - Surface finish change from laminate to varnish

- Outer carton reduction

Reduction of Materials – Horlicks Bag-In-Box

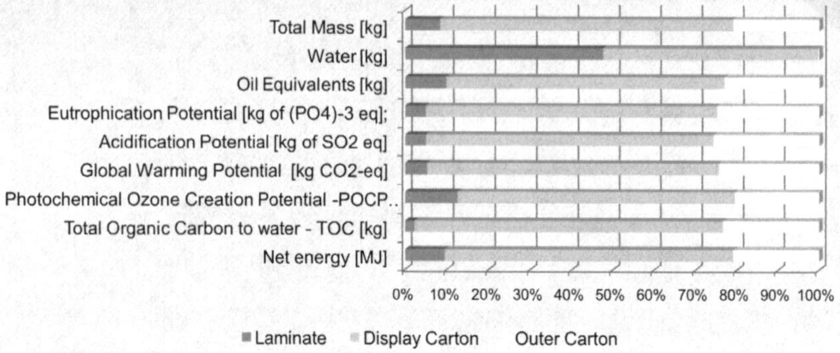

- Savings of
  - 65,210 GJ energy is equivalent to heating 15,890 homes for 1 month
  - 5008 tonnes CO_2eq is equivalent to taking 965 cars off the road
  - 2076 tonnes water is equivalent to 41,517 people's daily water needs
  - 261 tonnes carton is equivalent to 4348 trees

Glass to PP Bottles

- IndiaStar Award for Excellence in Packaging Award winner

Iodex Plastic Container – Eco Friendly Packaging Category

- 85% Reduction in Material Consumption vs the earlier pack
- Light weight, optimal balance of mechanical properties and excellent surface finish
- Lower Carbon Footprint compared to glass and metal packs
- Full body transparent shrink sleeve was used as a tamper evident feature has special printing of micro text with translucent ink which is a covert feature to protect it from counterfeiting

- Lower impacts from reduced materials and transportation

Iodex Glass to Plastic Bottles

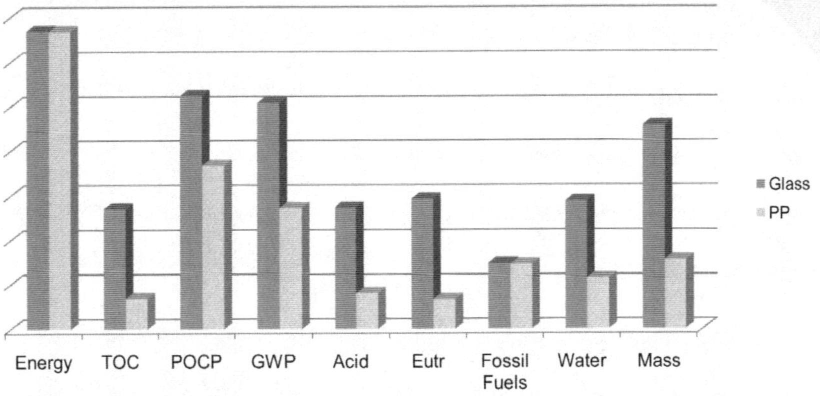

- Large decrease in mass of materials, but increase in raw material energy
- PP supplier closer
- Large decrease in energy for transport of raw materials from supplier and transport of product to distribution facilities

Horlicks Glass to Plastic Bottles

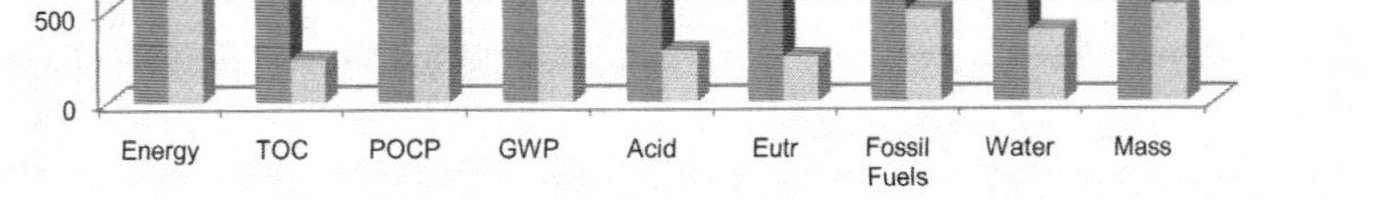

- Raw material energy very close
- 200g PP bottle raw material energy higher than for glass bottle
- Transport lower even though PP supplier farther
- Enough water saved for over 1 million people's daily water needs

LCA Impacts of 3 Insert Tray Options

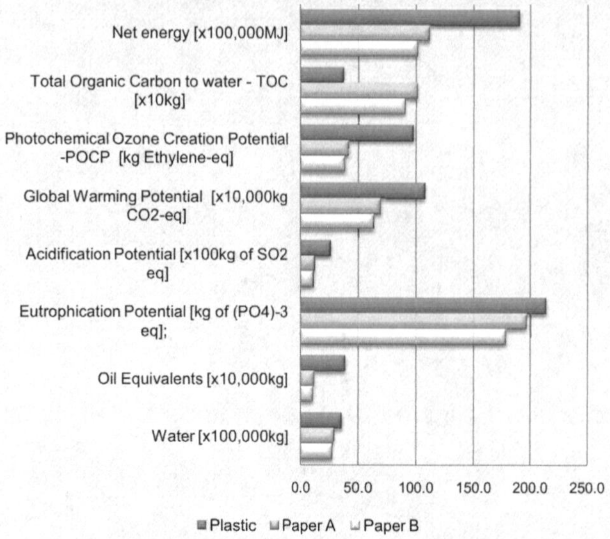

- Trade-offs – Paper B option lowest across all impacts except TOC

LCA Impacts of 3 Insert Tray Options

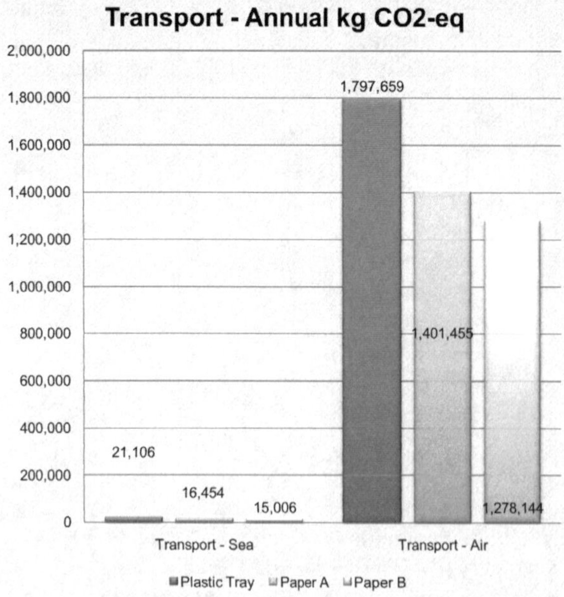

- Air shipping carbon footprint is 2 orders of magnitude larger than by sea

Bulk Packaging for Hospital Product

- Design
  - More product per carton – overall carton decrease
  - More product per corrugate – larger shipper
  - More product per pallet

- Overall Impacts
  - Impact decrease for carton
  - Impact increase for corrugate
  - Impact decrease for transport
  - Gave overall impact savings

Summary

- GSK is using life cycle approach to measuring environmental footprint of packaging

- Tools in place to help packaging engineers assess environmental footprint of packaging options

- Using these tools allow sustainability to be a part of the decision-making process when selecting new packaging designs

- Include as much of the life cycle as possible – Transport is important

- Environmental assessments should be done to assess individual scenarios

Acknowledgements

- Conchita Jimenez-Gonzalez
- Jennifer O'Donnell
- Liza Vernalls
- Martin Lale
- Linda Greener
- Richard Henderson
- Rebecca De Leeuwe
- Gregor Anderson
- Sukhdev Saini
- Richard Daw

gsk GlaxoSmithKline

Metrics Definitions

- Total Organic Carbon (TOC)
  - Amount of carbon present in organic matter [unit: kg]
- Photochemical Ozone Creation Potential (POCP)
  - The potential contribution of a material to smog formation. [unit: Ethylene equivalents]
- Global Warming Potential (GHG)
  - Any gas that absorbs infra-red radiation in the atmosphere. [units: kg CO_2-equivalent]
- Acidification Potential
  - The potential contribution of a material to increasing acidity in the environment. Acid gasses are generally normalized to the acid potential of gaseous sulfur dioxide. [unit: $(SO_2)^{-2}$ equivalents]
- Eutrophication Potential
  - The potential contribution of a material to excessive nutrient enrichment of the environment. Can cause increased algal growth leading to increases in oxygen demand and subsequent deprivation of oxygen for higher life forms. Materials that can lead to eutrophication are generally normalised to the nutrient potential of phosphate. [unit: $(PO_4)^{-3}$ equivalents]
- Oil Equivalents
  - The amount of fossil fuels used

Green Packaging Guide - WRAP

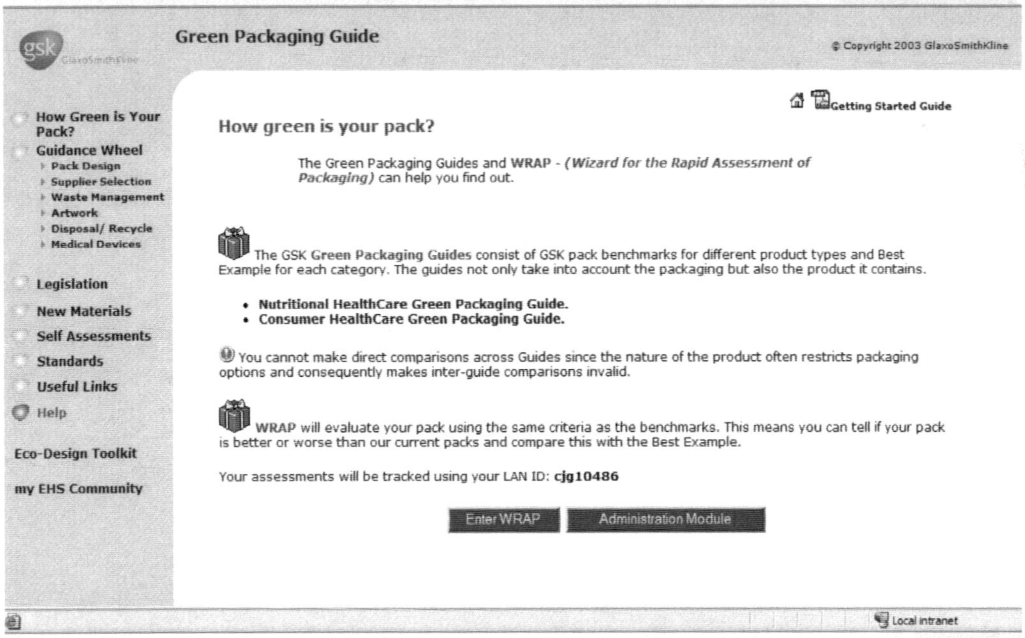

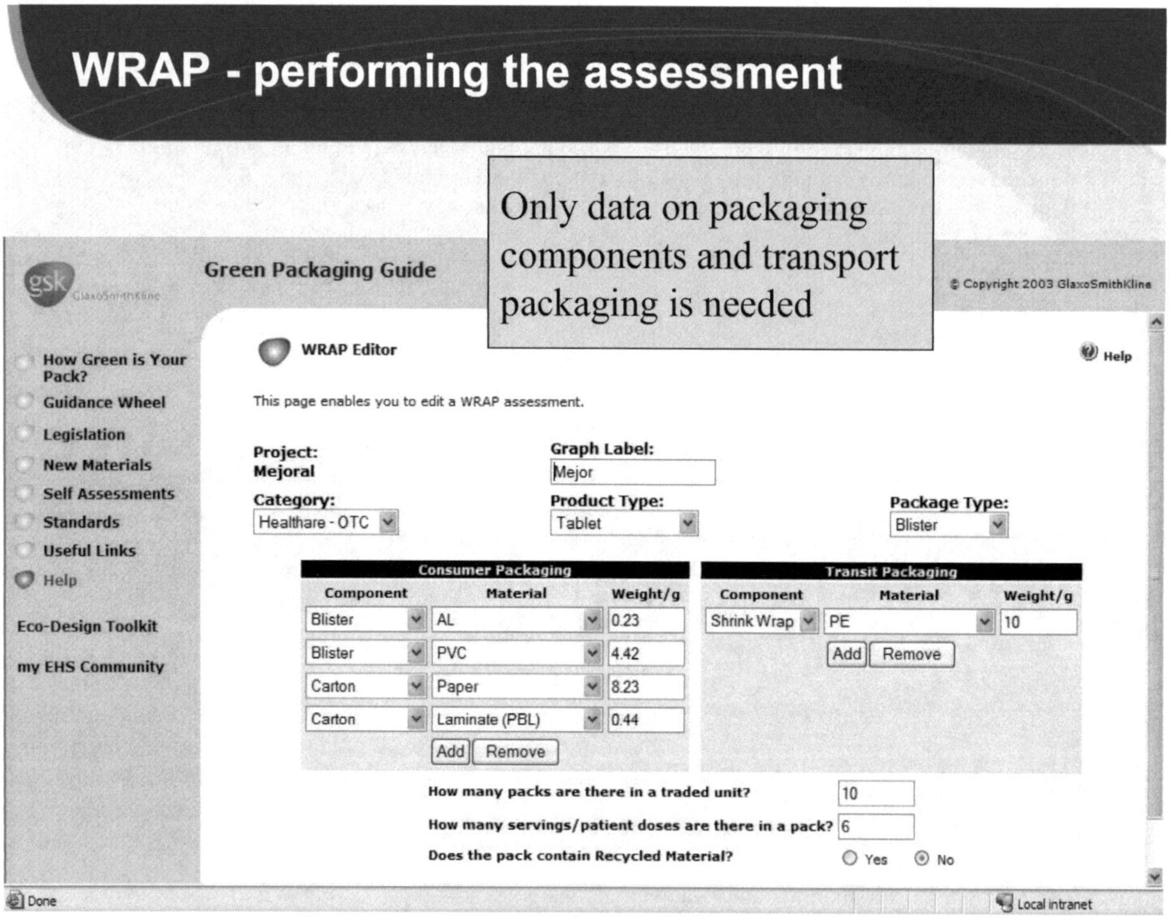
WRAP - performing the assessment
Only data on packaging components and transport packaging is needed
Green Packaging Guide
© Copyright 2003 GlaxoSmithKline
How Green is Your Pack?
Guidance Wheel
Legislation
New Materials
Self Assessments
Standards
Useful Links
Help
Eco-Design Toolkit
my EHS Community
WRAP Editor
Help
This page enables you to edit a WRAP assessment.
Project:
Mejoral
Graph Label:
Mejor
Category:
Healthare - OTC
Product Type:
Tablet
Package Type:
Blister
Consumer Packaging
Component
Material
Weight/g
Blister
AL
0.23
Blister
PVC
4.42
Carton
Paper
8.23
Carton
Laminate (PBL)
0.44
Add
Remove
Transit Packaging
Component
Material
Weight/g
Shrink Wrap
PE
10
Add
Remove
How many packs are there in a traded unit?
10
How many servings/patient doses are there in a pack?
6
Does the pack contain Recycled Material?
Yes
No
Done
Local intranet

AUTHOR INDEX

AIChE
120 Wall Street, FL 23
New York, NY 10005-4020

ISBN 978-1-5108-1736-4

CURRAN ASSOCIATES INC.
proceedings.com